ECOLOGY AND PHYTOGEOGRAPHY OF HIGH ALTITUDE PLANTS OF THE NORTHWEST HIMALAYA

Plate I. The Himalaya.

1. The Great Himalaya in the light of the rising full moon—a world where full many a flower bloom to blush unseen—the scene of the botanical studies described in this book.

2. *Primula rosea* in full bloom at the edge of a snow field, at an elevation of 4000 m on the Great Himalaya.

Ecology and Phytogeography of High Altitude Plants of the Northwest Himalaya

INTRODUCTION TO HIGH ALTITUDE BOTANY

M.S. MANI
M.A., D.Sc.
Emeritus Professor, School of Entomology,
St. John's College, Agra, India

LONDON
CHAPMAN AND HALL

First published 1978
by Chapman and Hall Ltd
11 New Fetter Lane, London EC4P 4EE

Printed in India by
Oxonian Press Private Ltd., Faridabad

ISBN 0 412 15710 1

Distributed in the U.S.A. by Halsted Press,
a Division of John Wiley & Sons, Inc., New York

Preface

This book is the outcome of over two decades of field explorations at high altitudes in the Northwest Himalaya. In the course of these explorations, undertaken mainly for the study of the high altitude insect life, extensive collections of plants were also made. In addition to collecting herbarium material, the plants were also photographed in their natural habitats. Observations on the general appearance of the plants, their abundance, associations of the species and distributional peculiarities were made. A four-month visit in the summer of 1963 to the USSR gave me the opportunity of personally exploring and collecting high altitude plants of the Central Tien Shan and the Alai-Pamirs.

The results of these entomological studies were summarized by me in my earlier books *Introduction to High Altitude Entomology* (1962) and *Ecology and Biogeography of High Altitude Insects* (1968). In a recent book, *Fundamentals of High Altitude Biology* (1974), I have also given an outline of the general biology of other organisms, including plants and man, at high altitudes. In this book, an attempt is made to summarize the salient results of observations on the ecology and phytogeography of the high altitude plants of the Northwest Himalaya. I have also reviewed the knowledge of high altitude plants of the Pamirs from the Russian language literature and in the light of my own observations and personal discussions with eminent Russian botanists like the late Prof. E.P. Korovin, who spent almost a whole lifetime in the study of the botany of Middle Asia, especially the Pamirs.

The book is divided into six chapters. In the first chapter, the concept of high altitude plants is explained, with a brief account of the Himalaya and its general flora and vegetation, as introduction to the high altitude plants of the region. This is followed by a brief outline of the salient characters of high altitude environment and of the high altitude ecology of the Northwest Himalaya. The third chapter deals with the synecology of high altitude plants of the Northwest Himalaya. The major ecological types, the plant formations, ecological affinities, etc., are briefly discussed. The fourth chapter reviews the important genera and species of plants, autochthonous to the high altitude ecosystem in the Northwest Himalaya. Though a taxonomic treatment is not attempted, brief notes on the plants should serve to illustrate the special ecological and phytogeographical ideas developed in the remaining chapters. This chapter provides the basic material for the discussions on the vertical distribution of high altitude plants presented in the fifth chapter and on the phytogeography and evolution contained in the sixth chapter. The high altitude ecosystem is shown to be phytogeographically an integral part of the Pamiro-Tibetan Province of Middle Asia. The bibliography of 124 titles should prove valuable to future students. The book is illustrated by many original text-figures and photographs. The photographs included here have been selected out of hundreds of Kodachrome transparencies, taken by me in the field, with Leica camera, equipped with Elmar 135 mm lens and closeup extension bellows and Telyt 400 mm lens.

The collections of the high altitude plants, made by me in the Himalaya, were mostly identified by myself in the herbaria of the Forest Research Institute, Dehra Dun, India and of the Institutes of Botany of the Academy of Sciences in Tashkent, Frunze and Tbilisi. Some material was also donated and identified for me by various specialists.

In writing this book, my aim is to draw attention to a most interesting, but greatly neglected field of biology. As the main emphasis is on the concept of high altitude, developed by me during the past two decades, review of the somewhat extensive literature on alpine and tundra plants is not attempted here. Students of alpine floras may find in the ecology of the higher elevations of the Northwest Himalaya, much that seems unfamiliar and perhaps even contrary to their experience elsewhere. This region is still largely unexplored and most species of plants known at present are in urgent need of careful revision by competent taxonomists. There is almost unlimited scope for study of the ecology of many interesting species. It is hoped that this book, with all its imperfections, would provide a useful basis for further explorations and stimulate research at high altitudes.

January 1978 M.S. MANI

List of Illustrations

Figures

Plates

Contents

Chapter I

Introduction

1. WHAT ARE HIGH ALTITUDE PLANTS?

HIGH altitude plants are essentially mountain forms that grow exclusively at high elevations and are integral parts of the high altitude ecosystem [77]. The differentiation and evolution of these plants are events confined to mountains and they are thus autochthonous to high altitudes. They do not represent accidental colonisations by plants of the lowland biota. While exceedingly few eurytype plants extend from the plains upwards on the wooded mountain slopes and may even be found growing above the limits of the forest on high mountains, the typical high altitude plants are, with rare exceptions as winter weeds, never met with in the lowlands.

The high altitude plants may be recognized by certain peculiarities, which can be described as part of their ecological specializations for the high altitude environment and which serve to distinguish them from other related types in the lowlands. Nearly every group of plants like lichens, mosses, liverworts, ferns, dicotyledons and monocotyledons, has specialized representatives in the high altitude ecosystem. Some families and many interesting genera are exclusively high altitude plants.

There is much confusion in biology about what really constitutes "high altitude". The divergent views have been reviewed and the modern concept of high altitude has been explained in two recent publications [76, 77]. In accordance with this concept, by high altitude we mean in biology the elevated regions on high mountains, above the upper limits of closed forest. It may be described as essentially a region of open vegetation, above the timberline on mountains, where the general environment, the flora, vegetation and animal life are significantly different from those of the forest or prairie. It is a highly specialized and distinctive ecosystem, which is characterized by the absence of trees.* As trees and closed forests may, however,

*On Equatorial mountains such as, for example, Mt. Kilimanjaro and Mt. Kenya, some high altitude plants like *Lobelia*, *Senecio*, etc., though not trees in the strict sense,

be absent even at relatively low altitudes and indeed disappear even quite near sea level in high northern latitudes, it is necessary to emphasize that only above a forest-line altitude of 2500–3000 m above mean sea level significant differences in the environment and ecosystem appear and can be clearly correlated with the increase in altitude. As it will become evident further on, these altitude limits are by no means arbitrary, but have sound biological basis. All high altitude plants evolve and grow at elevations above these critical limits.

The timberline that marks the ecological transition from the closed forest to the open vegetation on mountains is not, however, a sharply defined line at all. It is on the other hand a more or less oscillating zone of altitude, ascending here and descending there, in which the summer mean atmospheric temperature maximum does not rise above 10°C. It has been shown by Köppen [65, 66] and others that timberline is the altitudinal belt of mean annual 10° isotherm of air temperature in the warmest month. At temperatures below this limit, prevailing above the timberline, the growth of trees is no more possible. Instead of the closed forest, we have an open vegetation, characterized by stunted shrubs, low growing herbs and grassy meadows. Detailed accounts of timberlines in general and on mountains in particular may be found in a number of contributions by authors like Aynomin & Gupta [4], Brockmann-Jerosch [13], Gams [39, 40], Haviland [53], Larsen [72], Löchert [73], Lundegårdh [74], Mani [75, 76, 77], Schweinfurth [104], Shaw [105, 106], Troll [116, 117] and Wardle [121]. From our point of view, it is merely necessary to recognize that the timberline on mountains is an ecological transition between two fundamentally different ecosystems viz., the lowland and high altitude ecosystems.

The elevated regions above the limits of closed forests on mountains are generally known as "alpine zones". We have shown recently [75, 77] that the so-called alpine zone is really only very small part of the high altitude ecosystem, particularly on the Himalaya. The term is indeed entirely inappropriate with reference to the high altitude regions of the Himalaya [75]. Its limitations have also been recognized by a number of other workers. Swan [112] recognizes, for example, on the Himalaya a low altitude alpine zone from a high altitude alpine zone and at elevations higher above these zones he recognizes an aeolian zone, embracing the so-called arctic zone of authors and extending up to the summits of the great Himalayan peaks. The vegetational zonation into the temperate, alpine and arctic zones, generally followed, is not only confusing but must also be considered

have often a general arborescent appearance (gigantism). Though the terms 'alpine zone' or 'alpine tundra' are common, it has already been shown by me in earlier publications that these are unsatisfactory in the case of the Himalaya. The term 'high altitude' is adopted here in accordance with the concepts developed by me in my earlier books [75, 76 and 77].

as most unfortunate in the context of high altitude biology [77], because it misleads and overlooks a number of basic facts. These well known zones are no doubt more or less valid in a general way as one proceeds from the tropics to the high northern latitudes in the plains, but are not wholly satisfactory on high mountains. When these zones are extended to mountains, there is risk of overlooking a most important ecological factor that sharply distinguishes the high altitude from the high latitude ecosystem. In applying these zones to mountains biologists fall into the error of over-emphasizing the role of only the atmospheric temperature, isolated from other ecological factors. *The high altitude is not only a progressively cold region but is primarily a region of gradually semi-attenuated atmosphere; this is by no means the case with high latitude (arctic) regions.* Greatly impressed by the apparent parallelism between the fall in atmospheric temperature at high

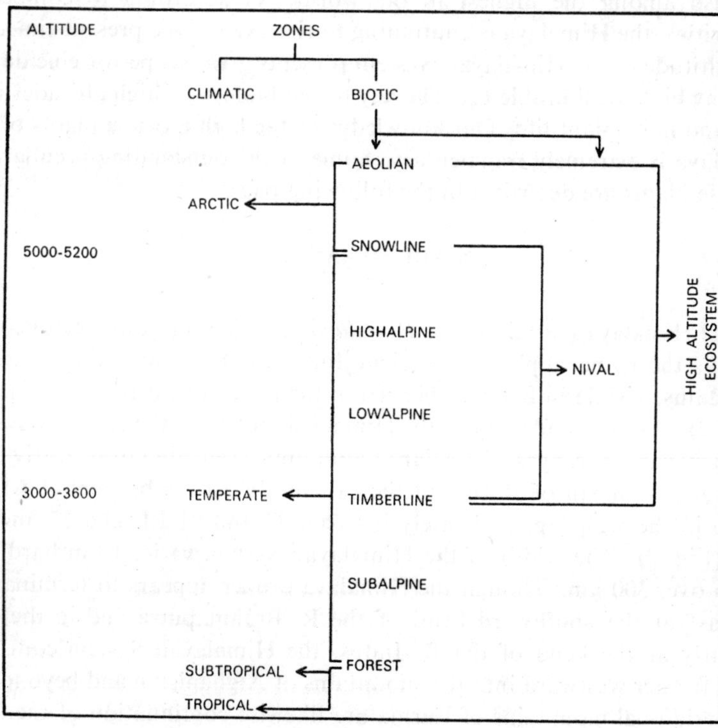

Fig. 1. Climatic and biotic zones on mountains comprising the high altitude ecosystem.

altitudes and high latitudes, biologists have entirely overlooked the basically different factors which underlie the atmospheric cold in the two cases. We have [77] recently drawn pointed attention to the fact that the temperate-arctic zones of north latitudes are basically *lowland* areas of moderate or low temperatures at dense atmosphere, but the high altitude is a region of low temperature of gradually thinner atmosphere (Fig. 6). For these reasons, and because of the great significance of snow cover at high altitudes for organisms, we recognized the elevated regions above the timberline on the Himalaya as *nival* regions [75], with special reference to insects. Although even in the case of plants the dependence on snow is by no means less significant, but because the plants are not so exclusively confined to the snow edge as insects are, we prefer to call them simply as *high altitude plants*. The high altitude plants include the subalpine, low and high alpine, nival, aeolian and arctic zone species. Defined in this way, by high altitude we mean the whole of the elevated regions above the timberline to the uppermost limits of possibility of life at present on the Himalaya (Fig. 1). The high altitude regions of the Himalaya are not only the most extensive, but also among the highest in the world. As according to competent authorities, the Himalaya is continuing to rise even at the present time, the high altitude of the Himalayan System presents great scope for elucidation of many biological problems. The key to the biology of high altitudes is to be found in its plant life. Our knowledge of the high altitude plants of the Himalaya is extremely fragmentary. Some of the outstanding peculiarities of these plants are described in the following pages.

2. THE HIMALAYA

The Himalaya (certainly not *Himalayas* as often but quite erroneously spelt) is the name applied in ancient India to the great snowy range of mountains, visible in fine weather, from the plains of north India (Pl. I, 1 & IV, 2). As now understood, the Himalaya embraces the complex system of nearly parallel ranges of Tertiary mountains, stretching over nearly 3000 km from the north of Burma in the east to almost the borders of Afghanistan in the west (approximately between 72 and 91°EL and 27 and 36° NL) (Fig. 2). The width of the Himalayan System varies from hardly 80 km to over 300 km. Though the Himalaya proper appears to terminate in the east at the southward bend of the R. Brahmaputra and in the west similarly at the bend of the R. Indus, the Himalayan System continues much further westward into the mountains of Afghanistan and beyond and the meridional mountains of Burma are likewise continuation of the same system. While the Himalayan System rises nearly abruptly from the plains of north India, it is continued in the north as a great series of folds of the crust, of which the Kuen Lun represents the northernmost. Between the Kuen Lun and the Himalaya proper lies the high bleak arid and windswept Tibetan Plateau, at elevations of 4750–4880 m above mean sea level. Excellent accounts of the geography and geology of the Himalayan System may be found in Burrard & Hayden [16], de Terra [23], Flint [36], Gansser

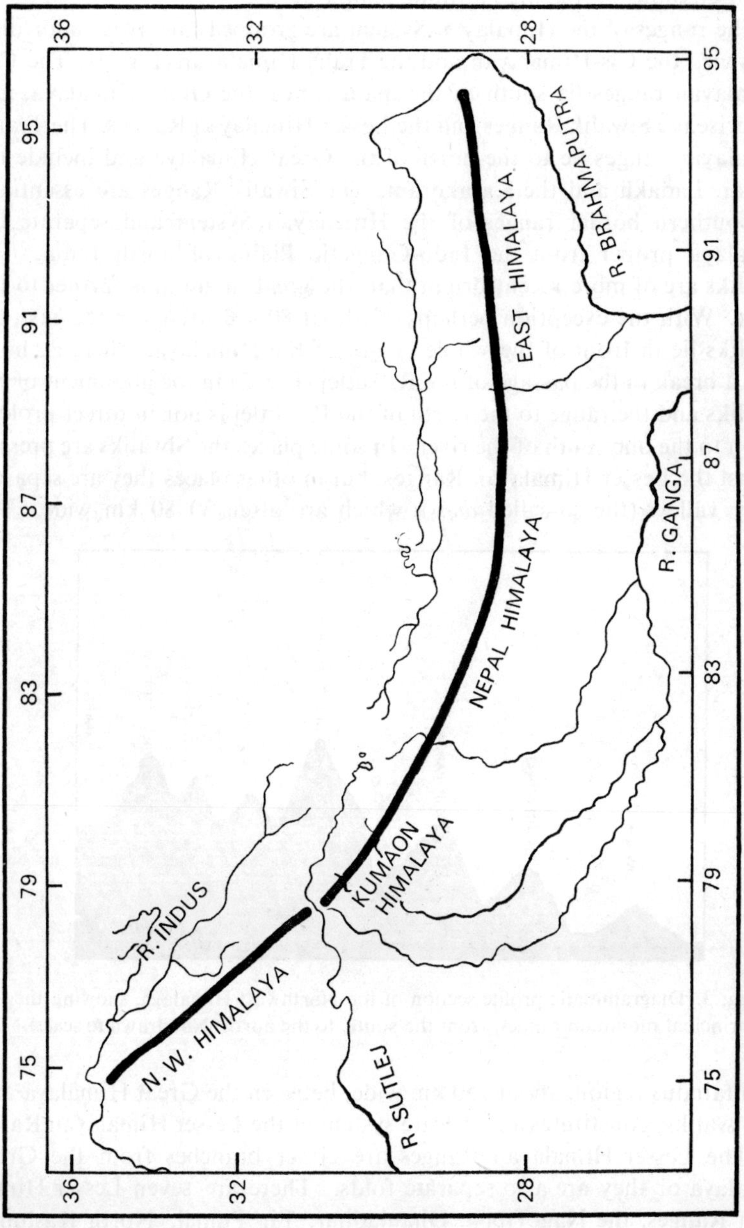

Fig. 2. Sketch-map of the Main Himalayan Range, showing the principal geographical sub-divisions.

[41], Heim [55], Heim & Gansser [56], Krishnan [70], Wadia [119] and Walker & Pascoe [120]. The brief outline that follows here will facilitate our discussions on the high altitude plants.

The ranges of the Himalayan System are grouped into two major divisions viz., the Cis-Himalayan and the Trans-Himalayan (Fig. 3). The Cis-Himalayan ranges lie south of the main range, the Great Himalaya, and comprise the Siwalik Ranges and the Lesser Himalayan Ranges. The Trans-Himalayan ranges lie to the north of the Great Himalaya and include the Zaskar, Ladakh and the Karakoram. The Siwalik Ranges are essentially the southern border ranges of the Himalayan System and separate the Himalaya proper from the Indo-Gangetic Plains of north India. The Siwaliks are of more recent origin than the great mountains further to the north. With the exception perhaps of about 80 km stretch in the east, the Siwaliks lie in front of the whole length of the Himalaya. There is, however, a break in the passage of the R. Sutlej (Fig. 2) in the alignment of the Siwaliks and the range to the north of the R. Sutlej is not in direct prolongation to the one south of the river. In some places the Siwaliks are pressed against the Lesser Himalayan Ranges, but in other places they are separated by valleys (the so-called *duns*), which are often 30–80 km wide. The

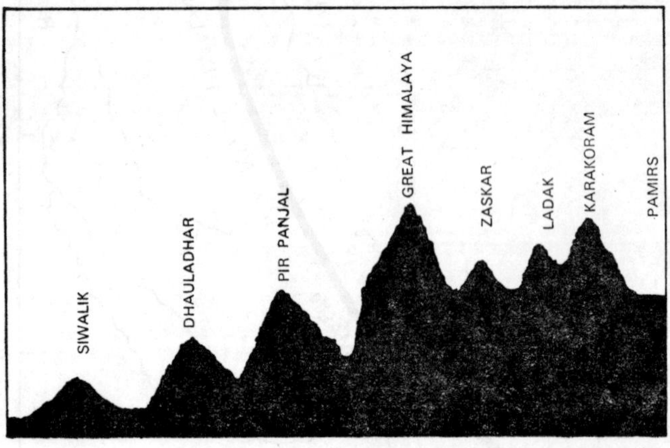

Fig. 3. Diagrammatic profile section of the Northwest Himalaya, showing the principal mountain ranges, from the south to the north (Not drawn to scale).

mountainous region, about 150 km wide, between the Great Himalaya and the Siwaliks, constitutes an intricate system of the Lesser Himalayan Ranges. The Lesser Himalayan Ranges are either branches from the Great Himalaya or they are also separate folds. There are seven Lesser Himalayan Ranges: the Nag Tibba, Dhauladhar, Pir Panjal, North Kashmir, Mahabharat, Mussurie and Ratan Pir. The Zaskar Range branches off from the Great Himalaya and the Ladakh Range extends from the region of Assam to Baltistan, though its continuity is not always evident. The Karakoram and the Hindu Kush Ranges are different sections of the same crustal fold that stretches from the southeast to the northwest; the eastern part of the fold is called Karakoram and the western part is the Hindu Kush. The Kailas Range is parallel to and lies about 70 km to the north of the Ladakh Range.

Current ideas on the structure of the Himalaya are modelled on the well known nappe theory of the Alps developed by de Terra [23]. The boundary faults, which were formerly interpreted as steeply dipping reversed faults that marked successive southern limits of mountain-building and the northern limits of the Tertiary sedimentation, are now explained as great thrust-planes. This border zone is now considered as the old surface of erosion, over which the older Himalayan formations were thrust and through the gaps of which they advanced in huge arch-shaped waves. The deposition of the Siwalik beds, of about 4575–6000 m thickness, seems to have been made possible by the tectonic downwarp.

The Himalaya is the result of a series of great orogenic movements, which were separated by periods of relative quiescence. The deformation of the crust is believed to have been initiated during the Upper Cretaceous and continued through the Middle Miocene, end of the Pliocene, the Pleistocene and Sub-Recent times. The Middle Miocene represents perhaps the time of maximum uplift, when the great masses of granites were intruded into the axial region of the Great Himalaya. The succession of the mountain ranges from the orogenic activity that transformed the Tethyan geosyncline is marked by three major phases: i) the elevation of the central axis of the ancient crystalline and sedimentary rocks during the Oligocene, ii) the Miocene movement that folded the sediments, and iii) the Post-Pliocene phase that affected the Miocene-Pleistocene Siwaliks sediments and which has not ceased yet. There has been a definitive southward drift of the orogenic activity, welding the successive belts of geosynclinal sediments onto the Middle Asiatic core.*

The areas that are now the Himalaya and Tibet were covered, during the middle of the Tertiary times, by an extension of the Tethys Sea, in which deposition of sediments had continued for a vast period of time. The Tethys Sea separated Eurasia from the Gondwana landmass. There were

*There is much confusion in English literature regarding Central Asia and Middle Asia. Most writers speak of Central Asia while actually referring to Middle Asia. It is necessary to define precisely the correct limits: Middle Asia comprises the Soviet Socialist Republics of Turkmenistan, Uzbekistan, Tadjikistan and Kirghiz (the Turkestan of old), the southern Kazakstan and north Afghanistan (to the north of the Hindu Kush Range); Central Asia includes the northern parts of Kazak SSR, the Mongol Peoples Republic, the western provinces of China (Sinkiang, Tsinghai), Tibet, parts of Kansu and the Inner Mongolia [76].

no high mountains, excepting the Altai in Eurasia and the Aravalli in the Gondwana region. The sedimentation from the Palaeozoic Era accumulated to a thickness of some 15200 m and was accompanied at the same time by a slow sinking of the sea bed. During a period of crustal movements, the floor of the sea began gradually to rise and came to be thrown into a series of long, parallel and wave-like folds. The crests of the waves were in course of time eroded by rain and weather and the rising land became much broken and highly irregular. Drainage basins thus came to be carved out of the flanks of the folds, giving rise to the river systems, composed of short transverse valleys. As the uplift of the mountains progressed, the troughs of the folds emerged to form a series of longitudinal valleys, at right angles to the transverse valleys, but parallel to the longitudinal axis of the crustal folds. A combination of concurrent uplift and erosion thus gave rise to the mountain system of the Himalaya and Middle Asia. The later folds came to be superimposed on the earlier folds, the arches were overturned, until almost horizontal and the whole region became greatly distorted and crumpled. The uplift of the mountains has, therefore, been brought about largely by horizontal pressure of the crust, acting in a meridional direction through long periods, continuing right down to the present. The wrinkling of the crust has taken diverse forms. The plateaus were wrinkled into ranges and the folded surfaces too were wrinkled and these wrinkled mountains have in their turn been corrugated to form smaller folds. The intervening troughs were filled with detritus from the mountains. In places where the stress has exceeded the breaking strain of the crust, the rocks have fractured and have greatly complicated the picture of structure of the Himalaya. Great parts of the crust have subsided and have also moved horizontally. Considerable molten material has been forced up from below in places of weakness and fracture and have in part absorbed the original sediments also. Thrusting is still believed to be in progress in the border regions and the vertical uplift of the mountains continues both in the outer and the main ranges. In the course of these changes, the Tertiary folding has wrapped itself in loops around the Gondwana Block. The influence of the Gondwana Peninsular Block on the alignment of the Himalaya is now generally recognized. The rocks have literally flowed around the tips of Peninsular wedges, anti-clockwise in Kashmir and clockwise in Assam.

Thrusting was perhaps not the only type of orogenic activity. Isostatic uplift seems also to have played an important part, perhaps at least in the last phase of the Himalayan uplift. The unloading, due to the shrinkage of the Pleistocene glaciers on the Himalaya, seems also to have brought isostatic uplift. The unloading brought about by the removal of vast quantities of the erosion products by the Himalayan rivers, should not also be overlooked. The rôle of the horizontal compression cannot of course be minimised; the high peaks like Mt. Everest, Kanchenjunga and Dhaulagiri, opposite the great foredeeps of the Gangetic Plains, are clearly evidence of the expression of a balance movement in the regions of the greatest exchange of load. The Himalayan compression is not simply an expression of the outward creep from Middle Asia, but largely also due to the extensive under-thrusting from the ocean floors and a definitive northward drive of old blocks from the south.

3. THE NORTHWEST HIMALAYA

It is obvious that a series of mountain ranges, as large and as long as the Himalayan System, should be divided, not merely as a matter of convenience but also on geographical and ecological grounds, into a number of divisions. The major geographical divisions of the Himalaya are as follows (Fig. 2): 1) the Eastern or Assam Himalaya, 2) the Central or the Nepal Himalaya, 3) the Kumaon-Garhwal or the Western Himalaya and the Northwest Himalaya or the Punjab-Kashmir Himalaya.

The Eastern Himalaya is about 720 km long and lies between the bend of the R. Brahmaputra to the R. Teesta. The Nepal Himalaya is about 900 km long from the R. Teesta westwards to the R. Kali. Three natural regions of the Nepal Himalaya are generally known viz., the Eastern Division or the Kosi Basin, the Central Division is the Gandak Basin and the Western Division is the Karnali (Gogra) Basin. The Western Himalaya is much corrugated and extends for about 320 km between the R. Kali and the great defile of the R. Sutlej. The Northwest Himalaya lies to the west of the Sutlej defile.

While the rest of the divisions are to an extent artificial, the separation of the Northwest Himalaya from the rest of the Himalaya must be described as essentially natural. Geographically, geologically, ecologically and biogeographically, the defile of the R. Sutlej is a most important landmark. Although nearly all botanists like Hooker [59], Clarke [20], Chatterjee [18] and others place the east-west phytogeographical transition in the Himalaya between Nepal and Kumaon, we have already demonstrated that the true biogeographical transition is at the defile of the R. Sutlej [75, 76, 78]. To the east of this defile, the Himalaya forms a biogeographical unit, which is fundamentally distinct and different from the Northwest Himalaya to the west of the defile. Some of the outstanding differences between the Northwest Himalaya and the rest of the Himalaya have already been discussed in some detail [75] and may be briefly recapitulated here.

There is at the defile of the R. Sutlej an abrupt physical break in the trendline of the mountain chains. While east of it, the ranges are mostly east-west, the trend of the ranges on the west is more southeast to northwest. The rest of the Himalaya lies largely between 27 and 29° NL, the

Northwest Himalaya stretches from about 30 to nearly 36° NL and has thus a much greater width and embraces a relatively extensive Sub-Himalayan tract. West of the Sutlej there is also an abrupt fall in the mean elevation of the mountain ranges. While many river gorges cut the main range (the well known antecedent river system) east of the Sutlej defile, no rivers pierce the main range in the Northwest Himalaya. The main water parting in the part east of the Sutlej defile, between India and Tibet lies to the north of the Great Himalaya, but in the Northwest Himalaya the water parting actually coincides with the crestline of the Great Himalaya. The Northwest Himalaya is believed to have been uplifted subsequent to the rest of the Himalaya. The other peculiarities of the general ecology, biology and biogeography of the Northwest Himalaya have been described in earlier publications; in the following pages is given an account of the plant ecology and geography. The flora and vegetation of the Northwest Himalaya are not so well known as those of the Eastern Himalaya. The Northwest Himalaya is remarkable for its ecology and phytogeography and presents many unsolved problems of great interest. As is shown in another chapter, the Sutlej defile, while marking an abrupt physical break, represents ecologically and phytogeographically a region in gradual east-west transition.

4. THE GENERAL ECOLOGY AND VEGETATION OF THE HIMALAYA (Pl. II–VI)

The outstanding peculiarities of the general ecology and vegetation of the Himalaya may be traced to its enormous massiveness, the great elevations of the mountain ranges, their trendlines, their locations in the middle of a vast continental mass, their Tertiary orogeny, the Pleistocene glaciations and the persistence of the uplift right into the recent times. Although situated only a few degrees north of the torrid zone, actually in the middle latitudes, owing to its enormous size and elevation, we find on the Himalaya an almost complete transition from the tropical to arctic conditions of climate.

Extending almost 3000 km from the east to west, it must be obvious that the ecological conditions and vegetation must differ very profoundly in the extreme eastern and western ends. This difference is particularly significant because of the monsoon climate to the south and the role that the Himalayan orogeny has apparently played in shaping this monsoon climate [78]. The general climatic conditions in the eastern parts of the Himalaya are almost semi-oceanic and largely humid tropical, but become more and more continental as we proceed westward. The general ecology of the Eastern Himalaya is greatly influenced by the Indian monsoon currents, but in the extreme western end the Himalaya, especially at high elevations, is nearly free from their influence. While the altitudinal zonation of vegetation and animal life is largely determined by the atmospheric temperature gradient, the marked east-west gradation of the ecosystems in the Himalaya is partly the result of close relation between the Himalaya and the monsoon-rainfall climate of India. The Himalaya is conspicuously more densely forested in the eastern parts than in the west. In the east, the forest is essentially an extension of the Indo-Chinese and Malayan floras and the tropical elements have in part spread to eastern Tibet, north of the Himalaya, and westward along almost the entire length of the southern slopes, even as far as Kashmir (Fig. 47). The wooded slopes of the Himalaya form a belt of very variable width, between the Indo-Gangetic Plains and the timberline above.

The Himalayan forests have been explored botanically by great many workers from early times. A great deal is known particularly about the vegetation and flora of the Eastern and Central Himalaya. We owe our knowledge of the plants of the Himalayan forests to the immense labours of many eminent botanists like Hooker [59]. Among the numerous other botanists, who have enriched our knowledge of the plant life of the Himalaya, mention may be made of Aitchison [1], Baehni, Bonner & Vauthier [5], Banerji [6, 7], Burkhill [15], Collett [21], Dey *et al.* [24], Dudgeon *et al.* [27], Duthie [28], Gupta [49, 50, 51], Hara [52], Kashyap [61, 62], Kingdon-Ward [63], Kihara [64], Landon [71], Marquand [79], Nair [87], Nakao [88], Numata [90], Rao [97, 98], Rau [99–101], Royle [102], Schweinfurth [104], Smith [107], Smythe [108], Troll [116, 117] and Zimmermann [124]. It would be wholly outside the scope of our work to attempt even a brief review of some of these brilliant contributions. The following general outline of the general vegetation and altitudinal zonation of the Himalaya is intended merely to illustrate the marked east-west transition in the ecosystems, so as to emphasize the ecological distinctiveness of the Northwest Himalaya. The Himalaya in the east is humid tropical, but the Northwest Himalaya is essentially arid cold steppes, particularly at higher elevations (Fig. 4).

The altitudinal zonation of vegetation is widely supposed to match almost exactly the zonation by latitude from the equator towards the poles [57]. Both pass climatically from the tropical, subtropical, warm temperate to temperate and cold arctic conditions and plant types. We have already pointed out, however, the anomaly of this parallelism [77]. Van Steenis [118] has also recently shown that this parallelism is by no means strictly true. He has explained in particular that mountain floras show rather abrupt altitudinal breaks in composition and character, so that the latitudinal-altitudinal similarity is not valid everywhere. Within these limitations, we may consider in broad outline the general altitudinal succession of plants in the Himalaya.

1. The Himalaya East of the Sutlej Defile

The Eastern Himalaya

The Siwaliks and the Lesser Himalayan Ranges are densley forested, especially on their southern slopes. In the east, the foot of the Himalaya, from the northern fringe of the plains, ascending almost up to an elevation of 700–900 m, is a region of humid tropical, largely evergreen rain-forest, dominated by *Shorea robusta*, *Adina stipulata*, *Dalbergia*, *Dillenia speciosa*, *Bauhinia vahlii*, *Anogeissus*, *Litsea*, *Phoebe*, *Lagerstroemia* and *Terminalia*. This is succeeded, at elevations of 1500–1700 m, by broad-leaved zone, dominated by *Castanopsis indica*, *Schima wallichii*, *Ficus*, *Engelhardtia*, *Euonymus*, *Michelia* and *Quercus*. Immediately above this belt lies a zone of *Quercus incana*, ascending to about 2800 m. At elevations of 2500–2800 m there is a marked dominance of Lauraceae, *Rhododendron arboreum*, *Acer*, *Lyonia ovalifolia*, *Symplocos*, etc. Above this we find a mischwald belt of conifers-rhododendrons, from an elevation of 2500 m to the timberline, which is characterized by *Betula utilis*, *Sorbus*, *Gaultheria*, *Abies spectabilis*, *Tsuga dumosa*, etc. At elevations between 4000 and 5000 m we find bush rhododendrons, juniper, *Androsace*, *Arenaria*, *Saussurea*, etc. The species of plants of the subalpine and the lower alpine zones belong mostly to *Berberis*, *Cotoneaster*, *Lonicera*, *Angelica*, *Astragalus*, *Corydalis*, *Epilobium*, *Heracleum*, *Potentilla*, *Rheum*, *Primula*, *Sedum*, *Androsace*, *Anemone*, *Dia-*

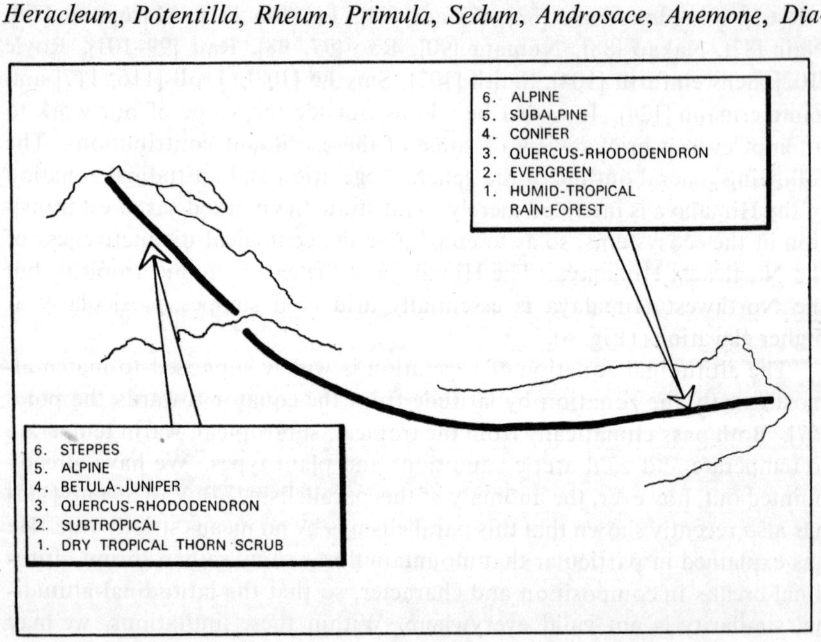

Fig. 4. The altitudinal vegetation succession in the Himalaya, showing in particular the marked differences in the east and west ends.

pensia, *Gentiana*, *Pedicularis*, *Lagotis*, *Carex*, *Juncus*, *Poa*, etc. In relatively drier localities grow *Arabis*, *Astragalus*, *Delphinium*, *Guldenstaedtia*, *Stellaria*, etc. The humid tropical and subtropical vegetation, at elevations of 600–1500 m in Sikkim is generally dominated by *Schima wallichii*, *Eugenia*, *Duabanga*, *Engelhardtia* and *Castanopsis*, with a great abundance of climbers, epiphytes, etc. The temperate zone in this region is characterized by the dominance of *Magnolia*, *Quercus*, *Acer* and Lauraceae. There is a marked preponderance of conifers at elevations of 2700–3600 m, especially forms like *Abies*, *Larix griffithii*, *Picea spinulosa* and *Taxus wallichiana*. The rhododendrons change from tree to bushes at 3600 m. The belt of 3000–3300 m is basically a zone of rhododendron-conifers, which become sparse higher above. We may summarize the succession as follows:

1. From the plains up to 700–900 m	Humid tropical rain-forest
2. From 700–900 m to 1500–1700 m	Evergreen forest
3. From 1500–1700 m to 2500–2800 m	*Quercus-Rhododendron* forest
4. From 2500–2800 m to 3600 m	Conifer forest
5. Between 3000 and 3600 m	Subalpine vegetation

The Central Himalaya

As already indicated, the Central or the Nepal Himalaya comprises an eastern, a middle and a western sector. The plants of the eastern and middle sectors have perhaps been best studied in the entire Himalaya. The eastern sector is basically a gradual thinning of the Sikkim-Bhutan flora of the Eastern Himalaya. In the middle sector at the foot of the hills, the forest is dominated by *Bombax*, *Ficus*, *Phoenix*, *Pandanus* and *Shorea robusta*. Above this comes the zone of *Castanopsis indica*, *Lithocarpus* and *Engelhardtia*, almost up to an elevation of 2500 m, with a great abundance of epiphytes. From an elevation of about 1500 m we start finding tree *Rhododendron*, *Quercus*, *Alnus*, *Prunus*, *Pyrus*, *Myrica*, scattered trees of *Schima wallichii* and some Lauraceae. In the upper limits of this zone, at an elevation of about 3300 m, start appearing *Quercus semecarpifolia*, *Abies*, *Betula utilis*, *Larix griffithii*, *Viburnum*, *Sorbus*, *Spiraea*, *Androsace*, *Gentiana*, *Pedicularis*, *Potentilla*, *Saxifraga*, *Primula*, etc. The timberline is characterized by *Rhododendron anthopogon*, *R. setosum*, *Juniperus squamosa*, etc. In the Kathmandu Valley the zonation begins with *Dendrocalamus*, *Arundinaria* and *Euphorbia royleana*, to be replaced higher up by *Shorea robusta* and *Pinus roxburghii*. Subtropical types like *Engelhardtia*, *Ficus* and arboriform *Euphorbia* reach up to elevations of 1500 m. At higher elevations of about 2000 m we come across *Rhododendron arboreum* and *Castanopsis-Quercus*. At elevations above 2500 m appear *Abis spectabilis*, *Tsuga dumosa*, *Rhododendron*, etc., reaching almost up to 3900 m. Above these limits lies the timberline vegetation, characterized by bushes of *Rhododendron barbatum*,

R. anthopogon, *R. nivale*, *Juniperus*, etc.*

The altitudinal succession in eastern Nepal may be summarized as follows:

1. Below 1200 m — *Shorea robusta*, *Lagerstroemia*, *Terminalia*, *Bombax*, *Bauhinia*, *Ficus*.
2. 1200–1900 m — *Schima wallichii*, *Castanopsis indica*, *C. tribuloides*, *Lithocarpus spicata*, *Alnus nepalensis*, *Myrica esculenta*, *Albizzia*, *Rhus*.
3. 1900–2500 m — *Quercus*, *Betula*, *Cylindrostachys*, *Cinnamomum*, *Machilus*, *Acer*, *Magnolia*, *Symplocos*, *Eurya*, *Rhododendron arboreum*.
4. 2500–3000 m — *Tsuga dumosa*, *Quercus semecarpifolia*, *Acer*, *Prunus*, Oleaceae, *Rhododendron* spp.
5. 3000–3900 m — *Abies spectabilis*, *Tsuga dumosa*, *Rhododendron arboreum*, *R. barbatum*, *Arundinaria*, *Viburnum*.
6. 3900–5000 m — *Juniperus wallichiana*, *Rhododendron campanulatum*, *R. barbatum*, *Betula utilis*, *Androsace*, *Cassiope*, *Saussurea*, *Sedum*, *Myricaria*, *Arenaria*.

The formations in the central sector of the Central Himalaya comprise tropical-subtropical formations, temperate formations, xerophile formations, steppes formations and alpine formations (Fig. 5). These formations may be briefly described.

A. TROPICAL-SUBTROPICAL FORMATIONS

1. Tropical formations up to elevations of 900–950 m: *Shorea robusta*, *Dillenia speciosa*, *Dalbergia*, *Acacia*, *Terminalia*, *Bauhinia*, *Anogeissus*, (*Bombax ceiba*, *Castanopsis indica*, *C. tribuloides*, *Holarhena antidysenterica*, etc., in pockets).

2. Subtropical humid forest between 900–950 m and 1900–2000 m: *Castanopsis indica*, *C. tribuloides*, *Schima wallichii*, *Daphniphyllum himalayensis*, *Albizzia procera*, *Terminalia belerica*, *Mahonia nepalensis*, *Berberis asiatica*, *Woodfordia fruticosa*, *Lantana*, *Rubus ellipticus*, *Colebrookia oppositifolia*, *Osbeckia stellata*, *Melastoma normale*, *Pogostemon plectranthoides*,

*Recently Numata and his collaborators have described the altitudinal vegetational zonation and the occurrence of deciduous broad-leaved forest in the cool-temperate zone of the extremely humid R. Arun Valley in the extreme eastern end of the Nepal (Central) Himalaya. The zones include *Shorea*-zone up to 1000 m, *Schima*-zone 1000–1900 m, *Quercus*-zone 1900–2500 m, *Acer*-zone 2500–2900 m, *Betula-Abies*-zone 2900–3750 m, *Rhododendron-Juniperus*-zone 3750–4500. They have included a comparative chart of the zones in other parts of the Himalaya. *Vide* Ohsawa, M., P.R. Sakhya & M. Numata, Mountaineering of Mt. Makalu II and Scientific Studies in Eastern Nepal 1971. Him. Committee of Chiba Univ. Japan, pp. 99–143 (1975); *Jap. J. Ecol.*, 23 (5): 218–228 (1973).

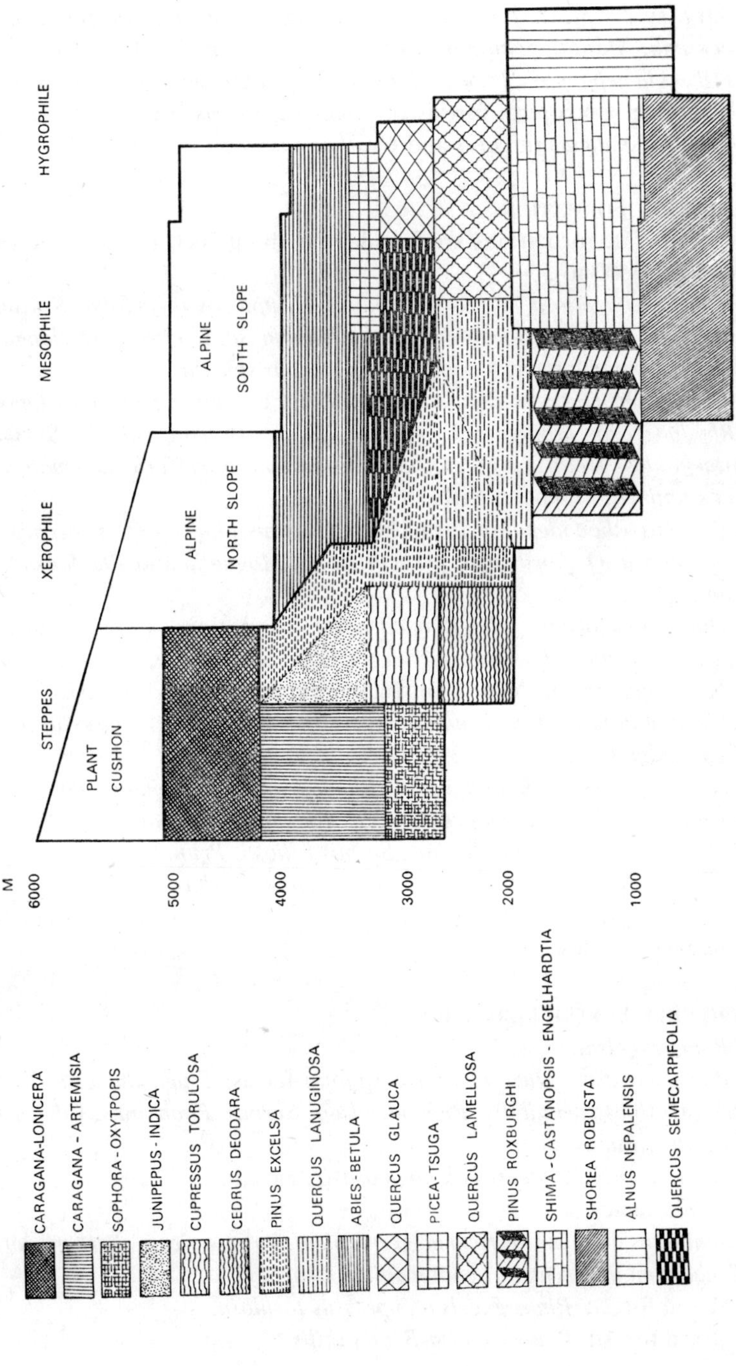

Fig. 5. A general scheme of horizontal and vertical succession of the major vegetation types in the middle sector of the Central Himalaya, based on the contributions of Dobremez & Jest.

Vitex negundo, Mussaenda roxburghii, Pteris aquilina, etc.

3. Dry subtropical: *Pinus roxburghii*-forest: *Engelhardtia spicata, E. colebrookiana, Mallotus philippinensis, Emblica officinalis, Wendlandia exserta, Osbeckia nepalensis, Phoenix humilis*, etc. (up to elevations of 1250 m).

4. Riverine subtropical forests of *Alnus nepalensis, Cedrela toona, Albizzia mollis, Pandanus furcatus* and *Cyathea spinulosa.*

B. TEMPERATE FORMATIONS

The temperate formations are found from about 2000 m up to the timberline (3000–3500 m).

1. *Quercus lamellosa* (hygrophile oaks), *Symplocos theaefolia, S. ramosissima, S. spicata, S. crataegoides, Litsea lanuginosa, L. citrata, L. elongata, L. umbrosa, Cinnamomum, Machilus, Lithocarpus spicata.*

2. Mesophile *Quercus-Rhododendron*-forest containing *Quercus lanuginosa, Rhododendron arboreum, Lyonia ovalifolia, Myrica esculenta, Symplocos crataegoides, Meliosma dilleniifolia, Litsea umbrosa, Coriaria nepalensis, Deutzia staminea, Mahonia, Berberis* and *Rhus.*

3. *Quercus-Rhododendron* formation: *Quercus glauca, Q. semecarpifolia, Q. lamellosa, Q. floribunda, Q. acutissima, Magnolia* and *Rhododendron arboreum.*

4. *Quercus semecarpifolia-Rhododendron* formation: *Acer, Magnolia, Carpinus, Viburnum, Euonymus, Ilex, Pieris, Sarcococca, Daphne, Ainslieaea, Arisaema intermedium* (from 2600–2700 m to 3200–3300 m).

5. *Picea smithiana-Tsuga dumosa* (2500–3500 m) with *Cupressus torulosa, Cedrus deodara.*

6. *Abies spectabilis-Betula utilis, Rhododendron campanulatum* 3300–3400 m to 3900–4000 m. *Juniperus indica, Rhododendron barbatum, Spiraea hypericifolia, Viburnum cotinifolium, Sorbus foliosa, Primula, Gentiana, Androsace, Rosa, Berberis, Cotoneaster, Meconopsis, Verbascum.*

7. *Abies spectabilis-Larix griffithii.*

8. *Bambusa*-facies.

C. XEROPHILE FOREST FORMATIONS

1. *Pinus excelsa.*

a. *Pinus excelsa—Quercus semecarpifolia* facies: *Tsuga dumosa, Taxus baccata* and *Abies spectabilis, Rhododendron, Sorbus, Euonymus, Viburnum, Prunus, Salix, Rubus.*

b. *Quercus* facies: *Quercus lanuginosa-Q. incana.*

c. *Epicea* facies.

d. *Juniperus indica*: *Myricaria germanica, Rosa sericea, Berberis aristata, Tanacetum nubigenum, Leontopodium stracheyi.*

e. Mixed forest: *Pinus excelsa-Cupressus torulosa.*

f. Mixed forest: *Pinus excelsa-Betula utilis.*

2. *Cedrus deodara-Cupressus torulosa.*
3. *Cupressus torulosa.*
4. *Juniperus indica.*

D. STEPPES FORMATIONS

1. *Sophora moorcraftiana-Oxytropis sericopetala* steppes: *Berberis mucifolia, Ephedra gerardiana, Lonicera hypoleuca, Incarvillea arguta.*

2. *Caragana gerardiana-Artemisia* steppes: *Artemisia* spp., *Rosa sericea, Lonicera rupicola, L. minutifolia, Spiraea bella, S. arcuata, Ribes orientalis, Berberis angulosa.*

3. *Caragana brevispinosa-Lonicera spinosa* steppes: *Juniperus squamosa, Potentilla fruticosa, Lonicera myrtillus, Hippophaë rhamnoides, Myricaria dahurica.*

E. ALPINE FORMATIONS

At elevations above 3900 m on the southern slopes and 4100 m on the northern slopes.

1. Alpine meadows (south) Gramineae, Cyperaceae, *Gentiana, Primula, Saxifraga, Pedicularis, Polygonum.*

2. Xerophile alpine (north) *Rhododendron anthopogon, R. lepidotum, R. lowndesii, Potentilla fruticosa, Juniperus squamata, Lonicera myrtillus.*

3. Discontinuous alpine vegetation of extreme altitudes: *Androsace sessiliflora, Thylacospermum rupifragum, Rhododendron nivale, Saxifraga* spp.

The western sector of the Central Himalaya is essentially an eastern extension of the Kumaon vegetation of the Western Himalaya. The succession from the plains closely follows that in the Western Himalaya, with the principal difference in the presence of *Pinus wallichiana, Picea smithiana, Tsuga dumosa* and *Cupressus torulosa* at elevations of about 2500–2800 m.

The Western Himalaya: Kumaon-Garhwal

The sub-montane area of the Siwaliks is characterized in this part of the Himalaya by *Shorea robusta*, which ascends up to an elevation of about 1000 m; this is basically a deciduous forest belt of *Lagerstroemia parviflora, Delbergia sissoo, Anogeissus latifolia* and *Terminalia.* At elevations between 1000 and 2000 m on the outer ranges, to the north of the Siwaliks, are found *Pinus roxburghii*, especially on the drier southern slopes. The upper limits of the *Pinus*-belt are characterized by associations of *Quercus incana* and *Rhododendron arboreum*, commencing from an elevation of about 1500 m, particularly well developed on sheltered slopes, with a great abundance of epiphytes, climbers like *Holboellia latifolia, Schisandra grandiflora* and *Vitis. Quercus incana* is replaced here by *Quercus dilatata* (= *Q. floribunda*) at elevations of 2000–2800 m, as a transition to *Quercus semecarpifolia*, which appears higher above. At elevations above 2800 m comes the zone of Quercus-Conifer, characterized by *Quercus semecarpifolia*, *Abies pindrow*, *Rhododendron arboreum*, *Euonymus*, *Viburnum*, *Taxus wallichiana*, etc. *Cedrus deodara* appears at elevations of 2000–3000 m. At elevations between 3000 and 4000 m is the timberline vegetation of *Betula utilis*, *Abies spectabilis* and *Rhododendron campanulatum*. Immediately above this are found scattered bushes of *Juniperus communis*, *J. wallichiana*, *Cotoneaster*, dwarf *Salix*, *Ribes* and the lower-alpine elements belonging to *Anemone*, *Ranunculus*, *Epilobium*, *Geranium*, *Gentiana*, *Polygonum*, *Potentilla*, etc. Bushes of *Rhododendron anthopogon* and *Berberis* may also be found at times at elevations up to 4800 m.

The zones of this part of the Himalaya are therefore: 1) Plains to the outer Siwaliks reaching to elevations of 1000 m; 2) *Pinus*-zone at elevations between 1000 and 2000 m; 3) *Rhododendron-Quercus* zone extending from about 1500 to 2800 m; and finally 4) *Betula-Juniperus-Abies*-zone at elevations of 2800–4000 m (100).

2. *The Himalaya West of the Sutlej Defile*

The Northwest Himalaya

The foot of the Himalaya is covered by subtropical, semi-desert vegetation, with dominant species like *Capparis decidua*, *C. spinosa*, *Ephedra*, *Pistacia*, *Cymbopogon*, *Enneapogon*, *Stipa*, etc., on the southern slopes. As we ascend from the plains of the Punjab on the Siwaliks and beyond on the foothills of the Lesser Himalaya, we come across a remarkable thorn-scrub vegetation, largely of *Acacia*, *Zizyphus*, succeeded by *Olea cuspidata*, *Dodonea viscosa*, *Pistacia integerrima*, *Punica granatum*, etc. At elevations of 900–1700 m we come across *Pinus roxburghii*, above which lies a belt of *Quercus*, *Rhododendron arboreum*, *Picea smithiana*, *Cedrus deodara*, *Abies pindrow*, etc. This is succeeded by *Betula utilis*, *Abies*, *Juniperus* and *Rohdodendron campanulatum* at elevations of 3000–3100 m, indicating timberline conditions. Extensive areas of arid mountain slopes are here covered by xerophytic arboriform *Euphorbia royleana*, particularly on the foothills of the eastern end of the Northwest Himalaya, and often extending deep within the Sutlej Valley.

The succession of zones from the south to the north, as we ascend the Northwest Himalaya, are thus: 1) The plains thorn-scrub forest, dry tropical vegetation; 2) Subtropical vegetation of the Siwaliks; 3) *Quercus-Rhododendron*-Conifer-zone; and 4) *Betula-Juniperus*-zone (timberline). Fig. 4 shows the striking differences in the altitudinal succession between the eastern and western ends of the Himalaya.

3. The Timberline and Snowline
(Pl. III–VI)

The timberline lies at elevations of 4200–4500 m in the extreme eastern end of the Himalaya, but descends to 2800–3000 m in the western extremity. The timberline is also higher on the southern slopes than on the northern and inner ranges. The timberline is often obscured locally by the deep penetration of tropical and subtropical vegetation, particularly in the river valleys, in the inner Himalaya, to the north of the Lesser Himalayan Ranges; this is, for example, remarkably clearly observed in the Sutlej Valley, where the timberline appears to lie somewhat higher than either to the east or west.

In the Northwest Himalaya, the timberline is generally much lower than in any other section of the Himalaya. Vast areas of the Northwest Himalaya really lie far above the upper limits of tree growth. Commencing from the northern slopes of the Pir Panjal Range, all the mountain ranges further north rise above the timberline altitude and are therefore characteristically barren. A detailed description of the plants of this belt lies wholly outside our scope.

The snowline, as commonly understood, would really convey no meaning under the conditions prevailing in the innermost Himalaya. Vast areas of rock are often completely snowfree even at very high elevations, largely because of the extreme atmospheric aridity in the region. At the same time, deep tongues of snow descend, however, to elevations below 5485 m. The mean value 5170 m usually given for the snowline in the Central Himalaya is also misleading. On the south slope of the Nepal mountains snow is usually abundant during June–September, even at elevations of 4575–4780 m, but on the inner ranges in the same region the mean atmospheric precipitation drops off abruptly, so that the northern gradient of scanty snowfall rises the snowline altitude. The term snowline is therefore strictly speaking, not a sharply defined altitudinal zone in the case of the Himalaya, where enormous tongues of snow descend much below 5000 m, and extensive areas higher above, even at about 6900 m, are completely free of snow. The snowline is on the average about a thousand metres lower on the south slope than on the north slope of the Great Himalaya. In the Northwest Himalaya the snowline is higher than in the Eastern Himalaya. As we proceed westward from the Eastern Himalaya, there is a gradual increase in general atmospheric aridity and consequently the mean snowline also rises higher.

5. SUMMARY

This chapter opens with an outline of the concept of high altitude in biology, developed by the author during the past two decades. This is followed by a general account of the Himalaya, its flora and general vegetation.

The high altitude plants are defined as mountain autochthonous forms that occur exclusively at high altitudes, above the timberline altitudes of 3000–3600 m above mean sea level, in the alpine, high alpine, nival, arctic and aeolian zones of earlier authors, and extending up to the highest limits of existence of plant life on mountains. The high altitude plants may be recognized by a distinctive combination of floristic, vegetational, ecological and physiognomic characters from those of other ecosystems.

The Himalaya is a system of parallel Tertiary mountain ranges, extending nearly east-west, and is in physical continuity with the Pamir Knot in the north. A brief outline of the geology and tectonics of the Himalaya, which is relevant to the main theme of the book, is presented.

As a preliminary to the discussion of the high altitude plants, the general vegetational and floristic characters of the Himalaya, with special reference to the altitudinal zonation and the east-west ecological-vegetational succession, are briefly reviewed. Particular attention is drawn to the striking geographical, geological, tectonic, hydrological, ecological and biogeographical distinctiveness of the Northwest Himalaya from the rest of the Himalaya. These facts lay the basic foundation for the ecological and phytogeographical discussions, which follow in subsequent chapters.

Cahpter II

Ecology of the High Altitudes of the Northwest Himalaya

1. THE HIGH ALTITUDE ENVIRONMENT IN GENERAL

THE outstanding characters of high altitude environments have recently been described, with special reference to the basic differences between the high altitude and lowland environments [77]. A brief outline of the fundamentals of high altitude environment should, however, prove useful in understanding the ecology of high altitudes of the Northwest Himalaya.

The high altitude environment is essentially an environment of progressively thinner atmosphere. The gravitational settling of the molecules of air near sea level results in a gradual fall in the density of air and with it also of the atmospheric pressure as the altitude increases (Fig. 6). It is not only the partial pressures of the component gases like nitrogen, oxygen and carbon dioxide, but also of the water vapour and the density of aerosol that progressively fall with the increase in altitude. The fall in the atmospheric pressure with the increase in altitude is associated with a whole series of complex chain events on high mountains, which contribute to the shaping of the high altitude environment.

Attention may be drawn here to some of these chain effects, in as far as they are important to plant life on mountains. The lowered density of air at high altitude results in increase of the transparency of the air, thus modifying the conditions of insolation, absorption and radiation and secondarily evaporation. Owing to the relative thinness of the air at high altitudes on mountains, only a small part of the solar radiation is directly absorbed by the air. At the same time, almost all of the terrestrial radiation is allowed to escape, so that the well known "green-house effect", with which we are familiar in the lowland regions, is nearly absent at high altitudes. The atmospheric temperature (in the shade) falls, therefore, with increase in altitude, in accordance with what is known as the "lapse rate". On high and massive mountains, the lapse rate of fall of air temperature with increase in altitude is, however, greatly modified by various complex and interacting factors like the general climate of the mountain region, the continentality of the mountain, latitude, massiveness, slope exposure, etc.

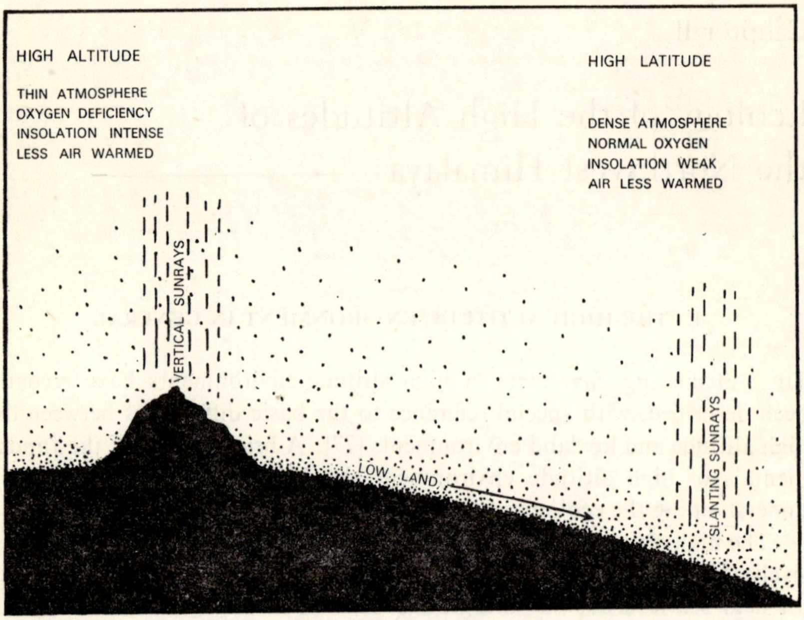

Fig. 6. The fundamental difference between the high altitude and high latitude ecosystems is traced to the gravitational settling of the air molecules near sea level. This results in dense atmosphere near sea level, both in the high boreal and austral latitudes, indicated by the denser shading. The progressive attenuation of air at high altitude is indicated by lighter shading. The atmospheric cold of the high altitude regions is due to factors fundamentally different from those operating in the case of the high boreal latitude cold region; the high altitude cold is related rather to the cold of the interplanetary space.

The altitude zone on mountains, at which the mean annual air temperature in the shade does not rise above 10°C is the timberline [65]. The altitude zone above the timberline, at which the air temperature similarly does not rise above the freezing point of water even in summer, is the snowline. Regions at altitudes higher than the snowline are characterised by the persistence of the snow cover throughout the year; the winter snow does not melt away completely.

The difference between the air temperatures in the shade and in direct sunshine, observed even near sea level and in all lowland areas, is greatly accentuated by the increassd air transparency at high altitude. The difference between the shade and sun temperatures thus tends to increase with increase in altitude. This difference may range, for example, from 35°C to as much as 73°C [75, 76, 77]. Regardless of the prevailing low atmospheric temperatures in the shade, most objects exposed to the direct sun at high altitudes on mountains, thus become warmed up far more rapidly than would be the case at sea level. These conditions largely influence the soil temperature, which is very important for the growth of roots and germination of seeds and minimizing freezing of the soil moisture.

The thin air and the scanty water vapour present in the high altitude atmosphere favour the rapid desiccation of bodies exposed to the air and to bright sunshine. It is generally found that the mean evaporating power of air at high altitudes, is mostly higher than should be expected at the prevailing air temperature.

The atmospheric cold, aridity, intense insolation and radiation and, snow cover, which are the chain effects of progressive attenuation of air with rise in altitude, are the major environmental factors for plants. The edaphic factors are largely of secondary importance in the high altitude ecosystem and do not necessarily determine the specific high altitude characters of the vegetation. It must, however, be pointed out that atmospheric cold, by itself, is not an absolute limiting factor for plants at high altitudes. Many plants, including certain remarkable Phanerogams, flourish at surprisingly extremely low temperatures, in localities above the permanent snowline. It seems that for plants at least, even if not for insects so much, the deciding factors are soil temperatures and snow cover that provide the necessary moisture. The critical conditions seem to be whether snow can at all melt under intensive insolation, even if the air temperature may be very low. The snow cover, the intensity of insolation, the hours of sunshine, the presence of even scanty moisture in the soil and the soil temperature, influence in their combination, the activities of plants at high altitudes far more than the atmospheric temperature. While the condition of atmospheric temperature merely seems to exclude certain kinds of plants and prevent the growth of trees, numerous other factors are important in shaping the characteristic high altitude plant ecology.

As in all lowland ecosystems, even in the high altitude ecosystem, it is not individual ecological factors, more or less isolated from others, but really the complex combination of numerous interacting factors that influence plants. All the environmental factors are mutually interactive; independent action of any single factor is unknown in Nature. The effect of any individual factor on a plant is thus very profoundly modified by the action of a number of other closely interlocked and mutually adjusting factors. Not only does each environmental factor act separately on the high altitude plant, but the sum total of all the factors complex is by itself, in its turn, a factor. In the effective factor complex of the high altitude environment, each factor exerts its own influence always in a definite interaction with every other factor. In all cases there is neither a simple alternation nor even a succession of the individual factors, but each factor works directly or indirectly on all others. Through this interaction it affects the plants in a most complex manner. All the environmental factors are thus closely and inseparably interlinked and intricately intertwined in a complex chain of cause and effect. In Nature there is, therefore, but a single, indivisible and complex factor, viz., the *high altitude environment*—an indivisible and integrated unit, in which we cannot isolate any factor element from the whole complex. The entire complex of the closely integrated and interlinked factors constitutes in reality a self-regulating dynamic system with the plants, which we recognize as the *high altitude ecosystem.*

Although these basic facts are well known, undue emphasis on the rôle of a single factor, considered in isolation from all others of the complex, has led to certain erroneous interpretation of the apparent parallelism between the atmospheric colds of the high altitude and high latitude. According to the well known Hopkin's bioclimatic law [57], a high altitude region of about 1500 m above mean sea level would correspond to the conditions of atmospheric temperature at sea level in the latitude 12° 30′ further to the north. There is also a general lag of seasonal events for each rise of elevation of about 122 m, corresponding to about one degree latitude in the north. Most biologists seem to have been so greatly impressed by this apparent parallelism that they have fallen into the serious error of equating the general climatic conditions at high altitudes on mountains with those prevailing at high north latitudes in the tundra and arctic regions. They have even identified the altitudinal biotic zonation with the biotic zonation

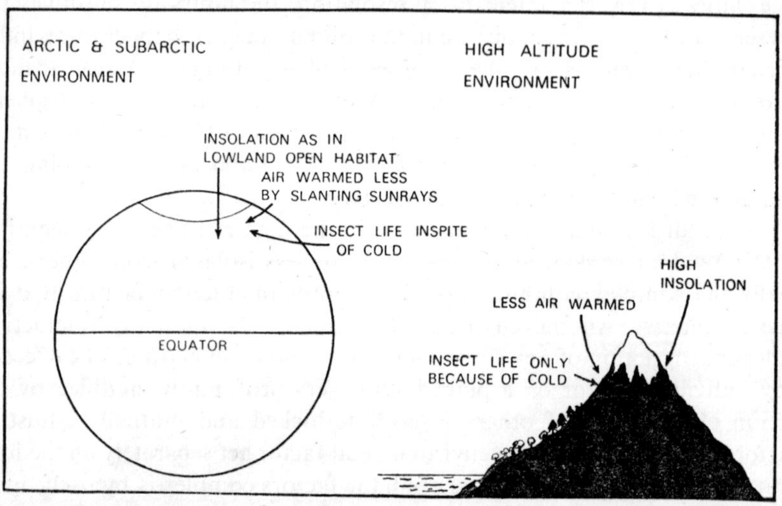

Fig. 7. High altitude versus Arctic-Subarctic environments. The fundamental differences in the ecology of atmospheric cold at high altitudes on mountains and near sea-level at high boreal latitudes and in the arctic regions.

at sea level, as for example Merriam's zones in North America [84], as we proceed from the equator to the North Pole.

The cold environments of high altitudes on mountains and of the north and south latitudes arise, however, from fundamentally different effects (Fig. 7) [77]. Their ecological values to organisms, not excluding plants, differ widely. The atmospheric cold of the subarctic, arctic and antarctic environments is conditioned not by the inadequate warming of a relatively thin or semi-attenuated atmosphere by the solar radiation, as is the case at high altitudes, but entirely by the difference in the angle of incidence of this radiation. The atmospheric cold of high latitudes is the result of relatively fewer sunrays incident on unit surface. At high altitude, on the other hand the atmospheric cold is the result of actually less air being warmed up, although there is in reality more sunrays per unit area of surface than at high latitudes. While certainly the mean atmospheric pressure and with it the density of the air are known to be influenced to some extent by the differences in latitude, even deep within the arctic or the antarctic regions, the atmosphere is by no means as thin as even on a moderately high mountain. High altitude cold is conditioned by precisely the same factors and is also more related to the cold of the aerospace environment and of the interplanetary space rather than to the arctic cold. In a sense the high altitude is a transition to the aerospace. It is, therefore, entirely erroneous to speak of the identity or even of parallelism between the high altitude environment and that of the tundra and arctic boreal regions. The intensity of insolation in high arctic latitudes is no more than that ordinarily observed in a lowland open habitat. As we have already pointed out, at high altitudes on mountains, insolation tends to become progressively very intense with increase of altitude. In this respect, the subarctic, arctic and antarctic environments are basically those of lowland open-vegetation habitats, under the atmospheric conditions typical of sea level. The environment of the high altitude on mountains is an open-vegetation habitat under more or less reduced atmosphere. Open habitats normally at sea-level atmosphere and open habitats of permanently thinner atmosphere cannot, (inspite of widespread misconceptions), evidently be ecologically identical or even similar. The properties, action, effects, modifications, interrelations, etc., of the environmental factors and of the organisms must be basically very different in the two cases. For these reasons, the familiar vegetational zonations of temperate, alpine and arctic are inappropriate in high altitude botany.

The high altitude environments of different mountains present characteristic differences. The environmental conditions, which may be attributed to altitude, are more or less profoundly modified by the massiveness of the mountain system, the trendline of the mountain ranges and the general climate of the surrounding lowland country. The high altitude environment, even at comparable elevations, is thus often different on different mountains, situated in different climatic regions of the world.

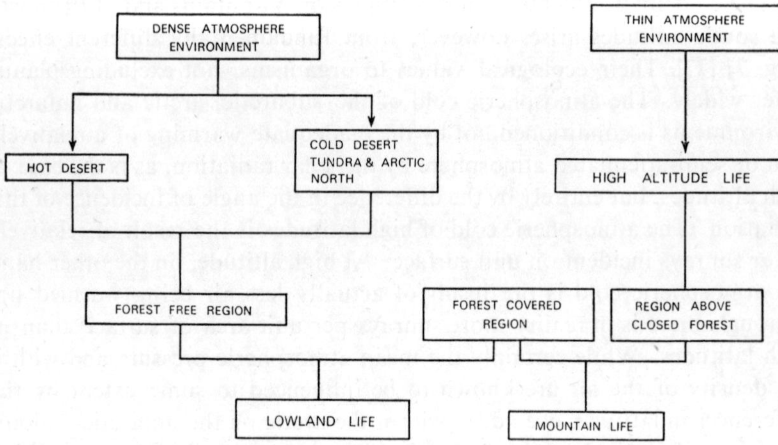

Fig. 8. High altitude versus lowland ecosystems. The open-vegetation ecosystem in forest-free areas under conditions of relatively dense atmosphere in lowland life, as met with in the tundra and arctic, is basically different from the open-vegetation ecosystem in thin atmosphere at high altitudes on mountains. Forest covered and forest-free areas of open-vegetation habitats occur both near sea level conditions and on mountains and constitute the basis for totally different ecosystems.

As is well known, tropical regions are generally characterized by the absence of sharply defined temperature-conditioned seasons. The north temperate and the north cold regions are large continental landmasses, with typically continental climate, characterised by sharp temperature-conditioned seasons. In the southern temperate and cold regions, we find relatively narrow meridional land areas or numerous scattered islands, surrounded by extensive oceans, so that the general climate is more oceanic and nearly isothermal conditions prevail the whole year round. These differences in the general climatic conditions of the lowlands very profoundly modify the high altitude environments of mountains situated in different regions (Fig. 7). The environment at high altitudes on equatorial and tropical mountains is remarkable for the dominance of diurnal temperature fluctuations over the seasonal fluctuations, so that there is no sharp difference between summer and winter. We find instead nocturnal sub-zero temperatures and frost every day, almost throughout the year at sufficiently high altitude. Nocturnal frost on tropical mountains may indeed occur even at elevations of 1500 m, depending on the general weather conditions, topography, vegetation, etc. Cloudy days prevent insolation and if they alternate with windless and starlit nights, the cold air caused by re-radiation is likely to stagnate and be cooled further, especially if the vegetation cover

Table I. Synopsis of the differences of high altitude environments in different climatic regions of the world (Fig. 8)

Character of environment	Mountains in north latitudes	Equatorial or tropical mountains	Mountains in southern latitudes
General climate	Continental; seasonal-thermic; seasonal temperature fluctuations are greater than the diurnal fluctuations	Diurnal-thermic; tropical; diurnal temperature fluctuations are greater than the seasonal fluctuation	Oceanic; isothermic; neither the seasonal nor the diurnal temperature fluctuations are pronounced
Atmospheric cold	Due both to the thin air and the slanting incidence of sunrays	Due only to the thin air	Due both to the thin air and the slanting incidence of sunrays
Insolation	Intense due to the thin air	Intense due to both the thin air and vertical sunrays	Due only to the thin air
Winter snow cover	Pronounced	No superabundant winter snow; nocturnal frost daily almost the whole year	Neither winter nor nocturnal snow very great

is low as turf, to give rise to frost. The high altitude environment on the mountains of the north temperate and boreal cold regions is characteristically seasonal-thermic, which marked differences in the mean winter and summer atmospheric temperatures. The high altitude environment of the mountains of the southern hemisphere is typically isothermic.

The characteristic differences of the high altitude environments in different climatic regions of the world are summarized in Table I.

2. THE HIGH ALTITUDE ENVIRONMENT OF THE NORTHWEST HIMALAYA

The peculiarities of the high altitude environment of the Northwest Himalaya may be briefly described as part of the characters of an essentially arid and cold steppes that has been elevated to high altitude conditions. These peculiarities are partly a result of the massiveness of the mountain region, the south-east and north-west trendlines of the mountain ranges and the physical continuity with the Pamirs. As index of the special features of the environment, we may consider here the atmospheric cold, snow cover and soil temperature in the Northwest Himalaya.

1. Atmospheric Tepmerature

In earlier publications [75, 76], we have given accounts of the atmospheric temperatures at different altitudes and have also discussed the variations observed. At elevations of about 3000–3900 m on the south slopes, the mean air temperature in the shade is about 8°C, but only 6°C on the north slope. In the altitude zone between 3900 and 5000 m, and in particular about 4200 m on the south slope, the mean air temperature during the year lies at 4.2° and 3.9°C on the north slope. At elevations above 5000 m and reaching nearly up to 5500–6000 m, the mean ranges from $-3°$ to $-6°$C, depending on the aspect and the mountain range. As may be expected, the lowest temperatures are observed during winter; the minimum is recorded sometime in January. The maximum is observed during July-August, when the summer may be said to be at its height. Fig. 9 summarizes the observations of monthly mean temperatures and precipitation at an elevation of about 3500 m on the south slope of the Great Himalaya in the Lahaul area. It must be understood that the mean temperatures do not often provide clues to the actual prevailing temperatures in any locality. At an elevation of about 4500 m on the south slope of the Great Himalaya the summer mean temperature may be 9.5°C and at an elevation of 5600 m it amounts to 7.0°C. On the south slope of the Pir Panjal range, at the timberline, the temperature during July has a mean value of 8.6°C, but during the winter drops to -8.2°C. On the Karakoram Range, at an elevation of about 7300 m, the air temperature is reported to be as low as −30.0°C in the shade. The annual temperature fluctuation depends naturally on altitude and mountain range; on the Pir Panjal Range it has been observed to amount to about 29°C at elevations of 3600–4200 m on the south slope. The diurnal temperature is greatly influenced by a variety of factors like cloudiness, wind, distance from the snow edge, etc., in addition to altitude, slope aspect and the mountain range, but is never larger than the annual variation. The daily maximum of air temperature is reached at about two in the afternoon during summer (June–July). After sunset there is a sharp fall in air temperature, but the minimum seems to be reached about early morning. In some localities, however, the minimum is reached much earlier during the night,

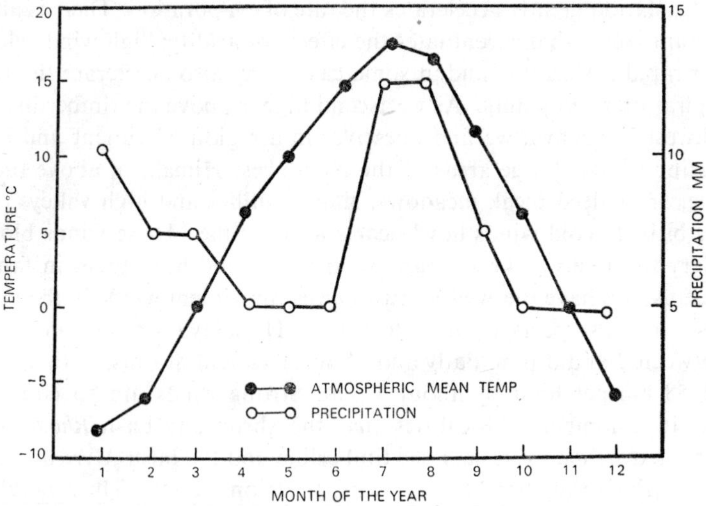

Fig. 9. The mean monthly atmospheric temperature and precipitation at an elevation of 3500 m on the Great Himalaya.

a few hours after sunset. This is particularly the case at higher elevations and in arid north slopes. These conditions are often very profoundly influenced by the prevailing winds, vicinity of snow, etc., but describe fairly accurately the average atmospheric conditions likely to be met with in the Northwest Himalaya. The difference between the shade and sun temperatures shows a decided tendency to be increasingly marked on the inner ranges and as the elevation increases. At an elevation of about 3900 m on the south slope of the Pir Panjal Range, the mean of this difference during May–June amounted to 39°C. On the 12th of June 1972 at about 11 a.m., with bright sunshine and light and steady wind, at an elevation of about 3690 m on the south slope of the Pir Panjal Range, the shade temperature was 8°C and the black-bulb thermometer registered 56°C.

2. *Atmospheric Aridity*

The elevated areas above the timberline in the Northwest Himalaya must be described as essentially the southward extension of the arid regions of high Middle Asia. It will help to appreciate the conditions, if we recollect that even the apparently arid localities at extreme elevations in the Eastern and Central Himalaya are more humid than the Northwest Himalaya. At an elevation of about 4500 m in the Northwest Himalaya, the air contains about 55% of the water vapour that it would hold at the same conditions of temperature in the plains. The mean evaporating power of the air is often two to eight times higher than inside the forest zone of the Himalaya. Even a slight rise of temperature under the influence of the intense insolation greatly accelerates the rate of evaporation. There is another important factor that accentuates the effects of aridity; high wind velocities favour rapid desiccation and in some cases may also accelerate the rate of transpiration from plants. As we ascend higher above the timberline in the Northwest Himalaya, we find ourselves in a region of violent and largely unceasing winds. Large areas of the Northwest Himalaya, above the timberline, are indeed bleak meadows, ridges, gullies and high valleys, where fierce, bitingly cold winds howl nearly all the time. These winds blow the powdery fresh snow like so many sand grains and heap them in fantastic shapes. As we have the well known *burane* or violent winds in the Pamirs, so also in most parts of the Northwest Himalaya strong winds spring up between 2 and 3 p.m. daily and blow as violent storms, with velocities of 54–58 km per hour by about 5 p.m. Strong winds are so constant a feature in a number of localities that the shrub and bush *Rhododendron* and stunted *Betula* even near the timberline and the bush *Myricaria* above the timberline show characteristic wind direction growth. The general severity of the winds increases as altitude rises and in the inner ranges. The effect of strong winds is to reduce the resultant insolation, lower the general air temperature and accelerate the evaporation of moisture from all exposed surfaces. These winds must be considered as one of the most important factors underlying the general aridity of the high altitudes; they also largely contribute to the peculiar physiognomy of the high altitude vegetation in many localities.

3. *Snow Cover*

The general aridity of the region naturally explains the scanty precipitation. The annual rainfall at Leh in Ladakh is reported, for example, to be no more than 75 mm and for several years in succession there is no rain at all. Precipitation at higher elevations is exclusively as snow. As may be expected, the mean annual snowfall increases as the altitude rises at first, but again diminishes at higher elevations, evidently as a result of the increasing atmospheric aridity with altitude. The Kumaon Himalaya receives more snowfall than the Northwest Himalaya. Although the south slopes generally receive a heavier snowfall than the northern slopes, the snow cover does not persist longer on the south, because of the greater insolation and exposure and also because of steepness of the south slopes. Depending upon diverse local factors, the winter snow may heap up to thickness of 20–30 metres. The first snowfall of the approaching winter may often occur even in September and generally continues in early spring next. The region between the timberline and the permanent snowline is generally under the cover of winter snow for about eight to nine months in the year. In contrast to the permanent snow, the winter snow is strictly seasonal and starts melting away with the coming of summer in May–June. The melting of the winter snow is progressively delayed as the altitude rises, as the arrival of the summer is also delayed as we ascend higher. It is thus not yet spring on the Great Himalaya when summer has already arrived on the Pir Panjal Range. These events are also delayed by a few days on the north slope when compared to the south slope of the same range. As the altitude rises above the timberline, not only does the summer arrive later, but is also correspondingly shorter. At elevations near the timberline on the Pir Panjal Range spring arrives usually by end of April or early May, so that the winter snow starts melting in mid May; by about end of July the summer is well advanced and nearly all the snow has melted away. At elevations between 3600 and 4200 m, spring arrives not until the end of May or perhaps even early June and mid summer is about the middle of August. The melting of snow starts significantly only in June and not all the snow has melted by end of August. On the Great Himalaya these events are further delayed; spring is delayed up to the end of May or early June and summer does not arrive even in early July. While near the timberline the summer lasts for about twenty weeks, it is hardly eight weeks in duration at elevations above 5500 m, so that not only is the metting of the snow much delayed, but also considerable snow cover persists well at the end of summer. As the melting of the winter snow progresses with the advancing summer, the edge of the receding snow ascends higher and higher. The altitudinal recession of the edge of the melting snow varies from about 200 to 240 m in about a fortnight or three weeks on the south slopes in Lahaul, in the height of summer. By about the middle of the third week of August summer is well advanced and may even be said to have ended and the edge of the snow has also receded nearly to elevations of 5000–5200 m.

It must of course be readily understood that snow is the only source of water for plants and all other organisms at high altitudes. The melt-water partly soaks into the ground, partly feeds the numerous streams and Himalayan rivers and is largely lost by evaporation during summer. It must not be concluded from the foregoing account that snow does not melt at elevations above the permanent snowline. While it is certainly true that all the snow does not melt away, it is also equally important to observe that even at elevations of about 6000 m, snow and glacier ice do melt under the influence of the intense daytime insolation in summer. There is therefore adequate film of moisture for certain class of plants, which thrive almost exclusively by virtue of such films. It thus happens that although the air temperatures in the shade are several degrees below the freezing point, strong insolation favours the presence of necessary moisture for plants on rock surface or in the scanty soil for pioneer species.

4. *Soil Temperature*

The peculiarities of soil temperature at different altitudes and the importance of soil temperature in high altitude ecology have been discussed by a number of workers. The available information has been briefly reviewed [75, 76]. With the melting of the winter snow in summer, the soil is exposed to the action of intense daytime insolation. The surface layers are readily warmed up on exposure to the sun and equally readily radiate away the heat when the sun sets. It is therefore found that the surface layers of the soil are subject to greater diurnal fluctuations in temperature than in lowland areas. It is also found that the soil temperature maximum fluctuates more than the minimum. At elevations of 3500–4000 m on the south slope, the difference between the surface soil temperature maximum and the mean air temperature during June is often 13.5°C, but only 5°C in the case of the minimum temperature.

As is well known, the soil temperature varies with the depth of the soil. At high altitudes, the diurnal fluctuations in the soil temperature during summer tend to become small with increase in depth, so that below a mean depth of about 15 cm, there is practically very little or no diurnal fluctuation. The temperature of the deep soil is, therefore, fairly uniform, irrespective of the fluctuations of air temperature and of the top soil. The deep soil temperature rises gradually with the advance of summer due to heat summation. The soil temperature varies appreciably with the distance from the snow edge. With increasing distance, the mean soil temperature shows an appreciable rise. At a distance of about 10 m from the snow edge the mean soil temperature at a depth of 20 cm on the south slope of the Pir Panjal during summer is about 2.5°C, but rises to nearly 7°C at a distance of 100 m from the snow edge. At a distance of over 20 m, the mean soil temperature at a depth of 20 cm is 9°C. The surface soil temperature rises more rapidly than the deep soil temperature with increasing distance from the snow edge. The deep soil temperature has profound influence on the penetration of melt-water, germination of seeds, growth of roots, rate of evaporation (transpiration) of water from soil, etc. The depth of penetration of soil by the melt-water and the rate of evaporation of soil moisture in their turn modify also the soil temperatures. The mean soil temperatures increase with altitude above 3900 m and continue to show increase up to almost 6000 m, but apparently diminish at higher elevations.

3. SUMMARY

This chapter has two major sections. The first section deals with the outstanding peculiarities of the ecology of high altitudes in general. It emphasises the basic differences of the high altitude and lowland ecosystems. Particular attention is drawn to the fact that the high altitude on mountains is basically a region of progressively thinner atmosphere, the chain effects of which are complex and mutually interacting environmental factors. An attempt is made to remove some of the widespread misconceptions about parallelism between the ecology of high altitudes and of the high boreal latitudes. It is stressed that as in all ecosystems, even in the high altitude ecosystem, it is not individual factors, in isolation from numerous others, that influence plants, but the combination of all the mutually interacting, interdependent factors like atmospheric cold and aridity, high air transparency, strong insolation and re-radiation, high glare and ultra-violet, excessive transpiration, strong winds, snow cover, soil temperature and moisture, brief and fleeting summer, rocky substratum, scanty soil, total absence of trees, etc. The high altitude environment is an indivisible and closely intergrated, self-regulating complex ecosystem, of which the plants constitute an integral part.

The second section deals briefly with the special features of the high altitude environment of the Northwest Himalaya. The peculiarities of shade and sun temperatures at different altitudes, the annual temperature fluctuations, soil temperatures and snow cover are briefly described. The Northwest Himalaya is a high altitude region of pronounced atmospheric cold and aridity, strong exposure and intense insolation, but adequate soil moisture derived from the melting of snow. While the Eastern Himalaya is a tropical high altitude ecosystem, the Northwest Himalaya is a continental elevated region of the middle latitudes, with steppes character.

Chapter III

Ecology of High Altitude Plants

1. INTRODUCTION

THE high altitude ecosystem, like all ecosystems, is a self-regulating, mutually interacting, closely integrated unit, composed of a great variety of habitats, communities, associations and formations of diverse genera and species of plants and other organisms. Not only the component units, but also the ecosystem as a whole is characterized by specific characters, which may be considered as part of the peculiarities correlated with the high altitude environment. These peculiarities serve to distinguish the member plants from those of the lowland ecosystems. Such peculiarities are superimposed on taxonomic and other characters, so that nearly all plants that belong to the ecosystem present more or less the same general and uniform habits. The general physiognomy of the plants of the high altitude ecosystem is quite characteristic and can be adequately described. The reader will find useful discussions on plants growing in localities of extreme ecological conditions, including the high altitude environment, by a number of authors like Löchert [73], on plant adaptations to cold-summer climate in arctic-alpine areas by Billings [8], effects of climatic and edaphic factors in general by Lundegårdth [74], general characters of the plants of the Alps by Braun-Blanquet [11], Gams [39], Schröter, [103] and others. All these authors have, however, apparently overlooked certain fundamental facts, peculiar to the high altitude ecosystems. Some of the peculiarities of high altitude plants in general have been briefly considered by me recently [77]. In this chapter is presented a brief outline of the basic peculiarities of high altitude plants of the Northwest Himalaya that serve to distinguish them not only from lowland plants but also from those of the Eastern and Central Himalaya.

Considered in broad outline of their specializations, the high altitude plants are essentially cryophytes and partly mesophytes or xerophytes. While a large group of animals, particularly the homoiotherms, are immediately affected by the increasing oxygen deficiency of the air, plants grow at extreme altitudes, where vertebrates and most other animals, except perhaps insects, cannot normally survive because of the inadequacy of oxygen. Plants occur, for example, at much higher altitudes than animals on the Himalaya [77]. The highest altitude at which an Angiosperm, *Christolea himalayensis*, has been found in the world is 6300 m on Mt. Kamet [100, 101]. Similarly cushions of flowering *Stellaria decumbens* were observed by Swan [111, 112] at an elevation of 6140 m on Mt. Makalu. We have given in Chapters IV and V examples of other Angiosperms, which flourish at elevations above 5500 m and almost up to 6000 m in the Northwest Himalaya. The highest elevation at which a Pteridophyte, *Cystopteris fragilis*, is known at present on the Himalaya is 5500 m [77]. Gymnosperms like *Ephedra* grow at elevations of 5200 m in the Northwest Himalaya. Moss and lichens flourish in great abundance, however, at much higher elevations than 6300 m on the Himalaya. A number of interesting lichens, like the yellow coloured incrustations of *Usnea*, Parmeriae and Gyrophorae, have been photographed by me on barren rock surface at elevations of 6300–6600 m on the Himalaya, where the intense summer insolation on the glacier ice produces a thin film of oozing moisture for brief periods. Reports of fungi and yeast in soil at elevations of 7600–8400 m on Mt. Everest may also be found in literature [112].

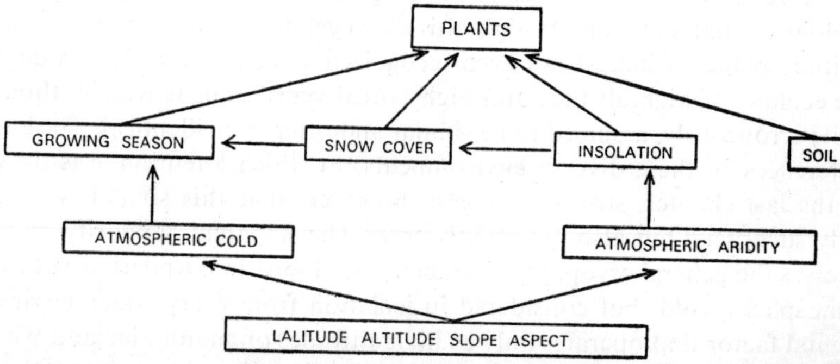

Fig. 10. The high altitude plants and limiting factors. The principal factors and their interactions, which largely determine the occurrence, abundance and ecological types of plants at high altitude. The progressive atmospheric cold and aridity, with increase in altitude on the Northwest Himalaya, rather than the increasing deficiency of oxygen in the air at high altitude, are the major determining factors in plant life. The upper altitudinal limits of plant life are influenced by the absence of soil moisture.

Although altitude by itself is apparently not a limiting factor, ecologically the plants are oriented at high altitudes primarily to the atmospheric cold and aridity rather than to the oxygen deficiency (Fig. 10). It seems that while low oxygen content of the high altitude air affects most animals, the low carbon dioxide content of air at extreme altitude and the failure of insolation to melt ice and snow are decisive factors in the ecology of high altitude plants. The deficiency of carbon dioxide in the air at high altitude should be expected to affect photosynthesis. It would appear that above a critical altitude, even though the daytime insolation during summer may provide sufficient moisture by melting snow and ice, the deficiency of carbon dioxide would inhibit extensive development of green vegetation. This conclusion would seem to be justified by the recent observation that the reduced amounts of carbohydrate reserves in roots and other underground parts of plants like *Polygonum bistortoides*, *Saxifraga rhomboidea* and *Calypteridium umbellatum* near the upper altitudinal limits, as compared with the values near their lower ranges, is to be correlated with diminished photosynthetic activity [14]. Quite apart from these facts, it is evident that the survival of high altitude plants at extreme elevations is not inspite of snow and glacier ice, but precisely because of their presence. Not only does the snow and ice provide all the moisture requirements for plants, but also serve as effective protective blanket to the dormant seeds and bud in winter.

2. THE HIGH ALTITUDE VEGETATION ECOLOGICALLY DISTINCTIVE

Considerable confusion seems to prevail at present regarding the general ecological characters of the high altitude vegetation. The ecology of high altitude plants is indeed not even recognized as distinctive. For example, the ecology of high altitude and high boreal vegetations is widely, though quite erroneously, assumed to be similar and often even identical. The basic differences in these diverse environments, to which attention was drawn in the last chapter, strongly suggest, however, that this similarity is not only superficial but also not strictly true. The apparent similarity merely stresses the general cryophyte character, developed in correlation with the atmospheric cold, but considered in isolation from every other environmental factor that operates only at high altitudes on mountains and which are totally absent in the boreal tundra and arctic regions. It must not, however, be overlooked that the boreal vegetation is composed essentially of only typically lowland plants, which have secondarily become specialized to the prevailing atmospheric cold. None of them is exposed to the chain effects of the atmospheric semi-attenuation of the high altitude environment.

Confusion extends even beyond the atmospheric cold and the related cryophyte character. The high altitude areas on continental mountains are usually classed, together with boreal latitudes, as cold deserts, primarily because of dominance of xerophyte characters. It must be clearly emphasized here that the origin and development of xerophytic characters of plants differ fundamentally in the hot desert, at high altitudes on mountains and in the high boreal latitudes of the tundra and arctic regions.

A little consideration will show that in the boreal tundra and arctic regions, the xerophytic character of plant is basically the result of scanty soil moisture for the roots, due to permafrost, and not entirely to atmospheric aridity. The permanently frozen ground and the insufficient thawing during the daytime in summer, because of poor insolation, and the frequent refreezing strongly inhibit the development of root system. The normal functioning of even the few roots is affected, so that the aerial parts of plants are largely starved of water and develop as peculiar xerophytes. In

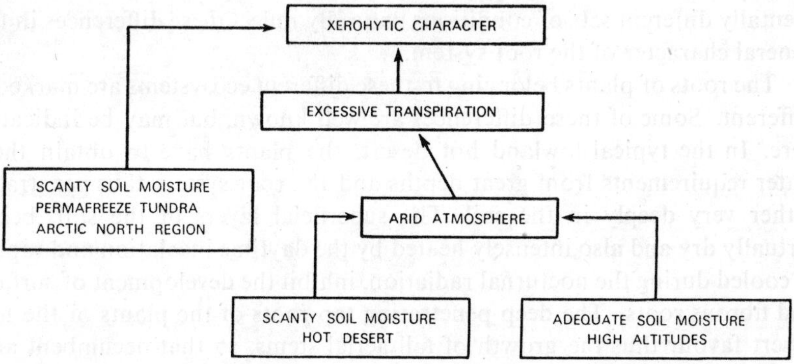

Fig. 11. Lowland and high altitude xerophytes. The fundamental difference between the xerophytes of lowland cold or hot deserts and of the high altitude ecosystems are to be traced to the basically different factors and plant reactions.

the hot desert, as is well known (Fig. 11), not only is the soil moisture practically absent, but the atmospheric aridity is also high. Insolation and radiation are intense and favour rapid desiccation. The roots fail, therefore, to absorb sufficient moisture and at the same time excessive transpiration from and desiccation of aerial parts are normal features. These combinations of conditions lead to pronounced xerophytic characters, which are basically different from those met with in the tundra regions. At high altitudes on mountains, in marked contrast to the hot desert and the tundra-arctic areas, there is generally adequate or even abundant soil moisture from the melting snow and ice [77]. The underground parts of nearly all plants at high altitudes are, therefore, characteristically succulent and the roots have usually considerable water available even at the soil surface. On the other hand, the arid atmosphere, the higher air transparency, the intense insolation and violent winds favour rapid transpiration from aerial parts, so that only these parts develop semi-xerophytic or xerophytic characters.

It is thus evident that the high altitude vegetation develops xerophytic characters primarily as a reaction to the atmospheric aridity, that of the tundra and arctic areas as a reaction to inadequate soil moisture and the vegetation of the hot desert as reaction to all-round lack of water, both in the soil and in the air [77]. Xerophytic characters develop, therefore, in plants under at least the following different sets of conditions: i) excessive transpiration and desiccation of aerial parts in arid atmosphere, high atmospheric temperature and scanty soil moisture; ii) excessive transpiration from aerial parts in arid atmosphere, pronounced atmospheric cold and intense insolation in direct sunshine, but with abundant soil moisture; iii) inadequate soil moisture because of frozen soil though exposed to atmospheric cold. The xerophytic characters developed under these fundamentally different sets of conditions naturally reflect these differences in the general character of the root system.

The roots of plants belonging to these different ecosystems are markedly different. Some of these differences are well known, but may be indicated here. In the typical lowland hot desert, the plants have to obtain their water requirements from great depths and the root system thus penetrates rather very deeply in the soil. The superficial layers of the soil, being virtually dry and also intensely heated by the daytime insolation and rapidly cooled during the nocturnal radiation, inhibit the development of surface and fibrous roots. The deep penetrating tap-roots of the plants of the hot desert favour thus the growth of tall aerial stems, so that decumbent and prostrate branches are on the whole uncommon. It is, therefore, true that in the hot desert trees are by no means excluded altogether; indeed many trees do grow wherever the deep soil moisture is available. In the cold desert of boreal tundra-arctic areas, the permanently frozen soil severely restricts and even completely prevents the development of deep penetrating tap-roots and this in turn reduces the growth of the aerial parts to stunted shrubs. At high altitude on mountains deep penetrating root system is precluded primarily because of scanty soil cover and the generally rocky substratum. Further, the abundance of water from melting snow even in the surface layers of soil makes it also entirely superfluous for the roots to penetrate deeply in the soil. We find, therefore, superficial fibrous roots the general character of high altitude plants. These conditions happen to be precisely the very factors that also preclude the aerial parts growing to shrub and tree heights. The superficial root system is closely correlated also with the short growing season and abundance of top soil moisture; the low atmospheric temperature, high atmospheric aridity are other contributory factors. In the lowland hot desert, nocturnal precipitation as dew is an important source of moisture for plants, so that the leaves absorb water during the night. There is also a great dominance of prickly and thorny plants. The xerophytes close down their stomata during the daytime and have succulent stems, phyllodes and other contrivances for storage of water. The hot desert sleeps during the day and may even appear dead, but revives at night. The flowers bloom largely at night and are predominantly white or yellow and are cross-pollinated by crepuscular or noctunal Lepidoptera. At high altitudes on mountains, on the other hand, there is no nocturnal absorption of moisture by leaves. With sunset the high altitude becomes dormant and all flowers bloom only during the hours of bright sunshine and are largely pollinated by wind or by diurnal Diptera and Lepidoptera. Aerial succulent parts for storage of water are totally unnecessary and prickly and thorny plants are by no means dominant in the high altitude ecosystem. High altitude vegetation is active only during the daytime and usually in summer, but sleeps at night and lies wholly dormant during the winter under the protective cover of snow. The high altitude xerophytes have thus many distinctive characters that sharply separate them from the lowland boreal and desert xerophytes. Korovin [69] has discussed in some detail the salient peculiarities of high altitude xerophytes, which he terms as mountain xerophytes in contradistinction to the lowland xerophytes. They differ from the lowland xerophytes not only in their genesis, but also in their general habits and appearance—the physiognomy of mountain xerophytes is distinctive. The general characters of cold mountain deserts, with special reference to the Central Tien Shan, are described by Kashkarov *et al.* [60A].

From the general dominance of semi-xerophytes and xerophytes, it must not, however, be concluded that all species of plants are xerophytes at high altitudes. It is true on the other hand that numerous species are extremely delicate herbaceous plants, with unbelievably tender leaves and flowers that wilt rapidly within a few minutes of being uprooted. There is in reality nearly every gradation from the strictly hygrophile to arid steppes type of species: 1) delicate hygrobiont annual herbs, 2) delicate hygrophile herbs which have the capacity of becoming dried up and of reviving when remoistened by melt-water, 3) semi-hygrophiles (mesophytes) of grassy meadows, 4) xerophiles and 5) xerobionts (euxerophytes).

While there are undoubted resemblances, the high altitude vegetation cannot be included under any of the lowland vegetation types known at present. It is on the other hand ecologically quite distinct from all others. The ecological distinctiveness of this vegetation may be described by the combination of complex ecological peculiarities of the component species, considered as a whole. The ecology of the high altitude plants of the North-west Himalaya is basically the ecology of plant life of an arid, cold, elevated continental mass of youthful topography. It differs markedly from the plant ecology of the Eastern and Central Himalaya and has many features common with the plant ecology of the Alai-Pamirs and the Central Tien Shan (*vide* Chapter VI).

3. THE GROWING SEASON

A striking feature of the general ecology of high altitude plants is the relatively short and fleeting growing season. The growing season, often also known as the vegetative season, is the period of active development and growth, of the germination of seeds, unfolding of buds, elongation of shoot, blooming of flowers and seed setting during spring and summer. The commencement of the growing season does not often synchronise with the coming of summer, but may be more or less delayed and may likewise fade away even before autumn. In most habitats, it is marked by the melting of the winter snow and the penetration of the soil by the melt-water. The growing season may be said to have properly arrived with the termination of dormancy of plants. It is a short, cyclic period that alternates annually with prolonged dormancy.

During the long winter months and well into the spring, the plants remain completely dormant under the protective cover of snow. Though nearly any part of a plant may become thus dormant, it is usually seeds, achenes, rootstock, underground stems with buds, etc., which survive the winter in the dormant state. Most of the dormant organs are usually densely clothed with dry imbricating scales, bases of petioles, dried leaf blades, etc., that serve as not only effective water-proofing but also as therma-insulators. Perennial aerial shoots have their buds similarly protected by dense woolly or cottony outgrowths or scales when they come to be buried under the winter snow.

When the winter snow melts in spring or early summer and the melt-water seeps through these protective coverings, the soil temperature also gradually rises. The seeds start germinating, buds revive and unfold and everything wakes up suddenly as if by magic—we have the growing season. The annuals sprout from the ground, grow remarkably rapidly, porduce a profusion of foliage, inflorescence or scape appear, and within a few weeks the colourful and massed flowers bloom everywhere on the mountain side. A week or so later, the flowers have all wilted, fruits appear and by the height of summer they ripen and the seeds have already matured. The leaves of the annuals have nearly all withered away, the seeds having been already scattered and safely covered over by debris and soil. In the aerial perennial parts, the flow of sap is already slowing down and there is a gradual cessation of all growth. Everywhere development and growth are ended and structural and physiological changes prepare the plants for the prolonged dormancy—the growing season is ended. The problem of dormancy is complex and largely obscure. We only know that dormancy is a necessary phase of a larger rhythm of development and growth, acquired in the course of the evolution of the ecosystem *pari passu* with the uplift of the Himalaya.

Although we know very little of the underlying factors, we may safely state that the decisive factor in terminating the dormancy of plants at high altitude is the melting of snow and the penetration by the melt-water of the dormant parts. The rise of the atmospheric or of the soil temperature does not seem to be alone responsible in breaking the dormancy in most high altitude plants. The exposure of the plants to prolonged cold, the long dormancy under snow and the return of moderately warm temperatures in early spring or summer are associated with a remarkably sudden outburst of intensive activity within the narrow strip of ground, freshly exposed from under the snow. The edge of melting snow gradually recedes higher and higher as the summer advances, successively exposing newer belts of dormant plants and leaving behind areas exposed earlier, drier and in which the plants have already set seeds. This march of revival of plant life is strikingly evident in the blooming *en masse*, producing a marvellous effect, of a massed carpet of green foliage and brightly coloured flowers on the mountain side, advancing higher and vanishing from below. This carpet seems to be bodily pulled upwards from the timberline to the snowline during summer.

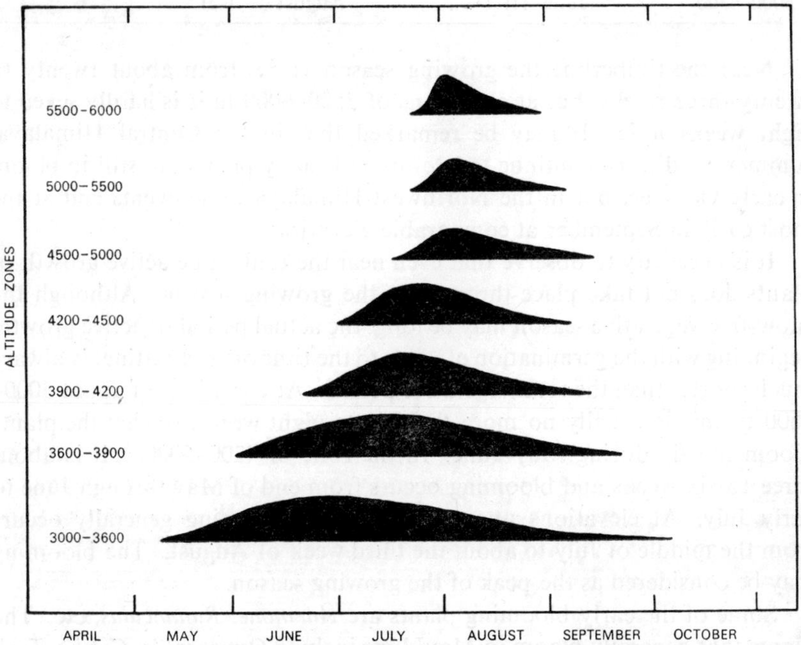

Fig. 12. The growing season in the high altitude ecosystem. The duration of the growing season is progressively shorter and it is also delayed as the altitude rises.

It may be recalled that in the Northwest Himalaya summer arrives later and is also shorter than in the Eastern and Central Himalaya. It has also been pointed out above that as the altitude rises, summer is progressively delayed and abbreviated. It must follow from these facts that the growing season must likewise be progressively delayed and must be extremely short at higher elevations. The growing season arrives somewhat later and also ends earlier on the north slope than on the south. The differences in the growing season in the Northwest and Eastern Himalaya have several ecological effects. The mean duration of the growing season at different altitudes in the Northwest Himalaya is summarized in Table II (Fig. 12).

Table II. The mean growing season at different altitudes in the Northwest Himalaya

Altitude	Start	End	Duration
2500–3000	April I–II week	October I–II	24 weeks
3000–3600	May I–II ,,	September IV	20 ,,
3900–4200	June I ,,	September III–IV	15 ,,
4500	June I–II ,,	September II–III	13 ,,
5000	June II–III ,,	August IV	10 ,,
5500	June III ,,	August II–III	8 ,,
6000–6300	June III–IV ,,	August I–II	6 ,,

Near the timberline the growing season varies from about twenty to twenty-three weeks, but at elevations of 5500–6000 m it is hardly seven to eight weeks long. It may be remarked that in the Central Himalaya, summer conditions continue to prevail and many plants are still in bloom in early October, but in the Northwest Himalaya these events end at the most early in September at comparable elevations.

It is necessary to observe that even near the timberline active growth of plants does not take place throughout the growing season. Although the snow-free vegetative season may be long, the actual period of active growth, beginning with the germination of seeds to the time of seed setting, is always much shorter than the total vegetative period. At elevations of about 3000–3600 m this is usually no more than six to eight weeks, so that the plants bloom mostly during May–June. In the zone of 4200–5000 m it is about three to six weeks and blooming occurs from end of May through June to early July. At elevations above 5500 m the blooming generally occurs from the middle of July to about the third week of August. The blooming may be considered as the peak of the growing season.

Some of the early blooming plants are *Anemone*, *Ranunculus*, etc. The plants that generally bloom in May–June include *Oxygraphis*, *Caltha*, *Trollius*, *Callianthemum*, *Thermopsis*, *Sisymbrium*, *Thlaspi*, *Braya*, *Erysimum*, *Geranium*, *Pimpinella*, *Pleurospermum*, most Compositae, *Primula*, *Gentiana*, *Swertia*, *Lindelofia*, etc. Among the species that bloom somewhat later in June–July are *Isopyrum*, *Parrya*, *Aquilegia*, *Cheiranthus*, *Barbarea*, *Arabis*, *Alyssum*, *Draba*, *Cochlearia*, *Chorispora*, *Saussurea*, *Artemisia*, *Myosotis*, *Dracocephalum*, etc. As may be readily understood, the same species blooms somewhat later at higher altitudes than in its lower altitudinal limits. Exceedingly few plants may be found in bloom during August–September and such species grow generally at elevations of 5500–6000 m and perhaps higher. *Parrya lanuginosa* and *Stellaria decumbens* bloom, for example, in September at elevations of about 6000 m. Swan [110, 111] has reported the blooming of *Selinum corticoides*, *Allardia glabra* (at 4725 m) and *Delphinium brunonianum* in October at elevations of about 5600 m on Mt. Makalu in Central Himalaya. *Parrya lanuginosa* was found by him in flower in October at an elevation of 6090 m in the same region. The time of blooming depends not only on the species, but on elevation and part of the Himalaya; blooming occurs later at higher elevations and in the Northwest Himalaya than in the Eastern Himalaya.

4. GENERAL PHYSIOGNOMY

The progressive abbreviation of the growing season with altitude is correlated with a number of other peculiarities of high altitude plants as chain effects and contribute to the distinctive general physiognomy of these plants.

The most striking feature that at once distinguishes the high altitude vegetation from that of the lower mountain slopes and of the forested plains is the complete absence of trees, large shrubs, lianas, climbers and epiphytes. Trees are, as we know, perennial plants that have one or rarely a few persistent and self-supporting main trunk, usually of considerable height (not to be confused with arboriform growth of certain species). Trees do not stand apart, but merge imperceptibly into shrubs, which are lower than trees and ordinarily have several branches arising directly from the ground or from close to the ground. Where trees grow, they very profoundly modify the entire environment, and provide optimal conditions for a great variety of weak-stemmed plants like twiners and climbers, besides supporting epiphytic Phanerogams and ferns on their branches. The total absence of trees and tall erect shrubs in the high altitude ecosystem accounts also for the absence of climbers. *Clematis orientalis*, which is a typical climber in the upper montane and subalpine zones, spreads like a mat on the ground at elevations of about 4267 m in the Northwest Himalaya.

Shrubs are generally confined to elevations immediately above the timberline or grow only in certain specially favourable localities at higher elevations. The shrubs generally found in the high altitude ecosystem of the Northwest Himalaya include *Lonicera*, *Myricaria*, *Hippophaë*, *Caragana*, *Astragalus*, *Gaultheria*, *Cassiope*, *Rhododendron anthopogon*, *R. campanulatum*, *Diapensia*, *Hyssopus*, *Perovskia*, etc. Most of these shrub species tend to be also characteristically dwarfed and tufted in the upper limits of their altitudinal range.

The greatest majority of plants grow close to the ground, remaining characteristically short and dwarfed, giving rise to densely tufted growths and semi-dome shaped or hemispherical cushions (as for example in *Thylacospermum*, *Macrotomia*, etc.) (Fig. 30) or polsters that spread on the ground. Most of them have an average maximum height of about 30–40 cm and only exceedingly rarely, species like *Iberidella* and *Morina coulteriana* ever exceed a height of 60–70 cm and are very exceptionally about a metre high. Some of the plants, which thus become conspicuously dwarfed at higher elevations include *Ranunculus*, *Oxygraphis*, *Trollius*, *Aconitum*, *Cimifuga*, *Stylophorum*, *Meconopsis*, *Corydalis*, *Parrya*, *Barbarea*, *Arabis*, *Alyssum*, *Cochlearia*, *Sisymbrium*, *Christolea*, *Braya*, *Capsella*, *Thlaspi*, *Viola*, *Stellaria*, *Sagina*, *Cousinia*, *Aster*, *Erigeron*, *Inula*, *Senecio*, *Oxytropis*, *Hieracium*, *Acantholimon*, *Calamintha*, *Eurotia*, *Axyris*, *Kochia*, etc. Tufted growth distinguishes a great number of plants like *Anemone*, *Caltha*, *Isopyrum*, *Callianthemum*, *Cheiranthus*, *Draba*, *Chorispora*, *Erysimum*, *Eutrema*, *Gypsophila*, *Silene*, *Lychnis*, *Cerastium*, *Arenaria*, *Sagina*, *Biebersteinia*, *Saxifraga*, *Leontopodium*, *Tanacetum*, *Saussurea*, etc. In these species a large number of short, annual branches arise each summer, close together from near the ground, on persistent underground perennial stems.

The absence of trees and large shrubs does not necessarily mean that all plants are ephemeral and annuals in the high altitude ecosystem. The high altitude environment favours, on the other hand, the abundant development of remarkable perennials. The growing season is no doubt not long enough for many plants to complete their full development within a year. They have, however, the capacity of continuing the development after an interval or even next summer. While the annuals do manage to complete the entire life cycle of vegetative growth and reproductive development of flowers and setting of seeds within the brief growing season, most high altitude plants are, however, typically perennials and require indeed several years to attain maturity. They become dormant at any time, without suffering serious injury if the conditions turn suboptimal even during the growing season. Development and growth may thus be interrupted any time during the summer and may not always be resumed the next summer. Although the plants continue to grow for several years, they remain characteristically low and close to the ground. Inspite of their small size, most of the high altitude plants are of great age. While the general size of the plants seems to diminish with increase in altitude, their longevity does not.

A basic difference between the general dominance of perennials in the lowland and high altitude ecosystems must be noted. In the lowlands there is a very marked dominance of aerial perennial parts of the plants like persistent trunks, but in the high altitude ecosystem the aerial parts of plants are, with some notable exceptions, rarely if ever perennial. The dominant perennials of the high altitude ecosystem have typically ephemeral or annual aerial parts that wither away at the end of each growing season.

They have on the other hand perennial underground stems, rhizomes, stolons, rootstocks, creeping and often also woody, long rooting runners, generally covered by thick layers of old leaf sheaths, expanded bases of petioles, etc., and sending out fibrous roots. These sprout forth annually, during the growing season, into prostrate or decumbent, or erect and dwarf or tufted aerial shoots that wither away after having set seeds. There is, therefore, a great preponderance of perennials or plants with perennial underground parts. Hardly one-sixth of the plants are annuals at extreme high elevations. Perennials or plants with perennial persistent parts belong to species of *Anemone*, *Adonis*, *Cochlearia*, *Bierbersteinia*, *Thermopsis*, *Ranunculus*, *Oxygraphis*, *Caltha*, *Trollius*, *Isopyrum*, *Aquilegia*, *Aconitum*, *Callianthemum*, *Cimifuga*, *Stylophorum*, *Meconopsis*, *Parrya*, *Barbarea*, *Arabis*, *Draba*, *Christolea*, *Cardamine*, *Eutrema*, *Cerastium*, *Stellaria*, *Arenaria*, *Thylacospermum*, *Myricaria*, *Geranium*, *Ononis*, *Caragana*, *Astragalus*, *Oxytropis*, *Spiraea*, *Geum*, *Fragaria*, *Potentilla*, *Saxifraga*, *Parnassia*, *Bupleurum*, *Pimpinella*, *Pleurospermum*, *Lonicera*, *Scabiosa*, *Leontopodium*, *Anaphalis*, *Tanacetum*, *Artemisia*, *Cremanthodium*, *Senecio*, *Echinops*, *Carduus*, *Saussurea*, *Jurinea*, *Crepis*, *Acantholimon*, *Diapensia*, *Androsace*, *Swertia*, *Cynoglossum*, *Lindelofia*, *Solenanthus*, *Paracaryum*, *Echinospermum*, *Eritrichium*, *Myosotis*, *Macrotomia*, *Scrophularia*, *Pedicularis*, *Lagotis*, *Marrubium*, *Kochia*, many species of *Polygonum*, *Aletris*, *Eremurus*, *Fritillaria*, *Cobresia*, *Carex* and *Stipa*. The annuals with perennial underground parts are typically species of *Sedum*, *Sempervivum*, *Valeriana*, *Primula*, *Arnebia*, *Picrorhiza*, *Veronica*, *Iris* and *Allium*. The typical annuals, which lack such perennial parts and yet flourish at extreme high altitudes include interesting species of *Delphinium*, *Hypecoum*, *Corydalis*, *Cheiranthus*, *Sisymbrium*, *Viola*, *Gypsophila*, *Silene*, *Lychnis*, *Sagina*, *Trifolium*, *Melilotus*, *Epilobium*, *Aster*, *Exacum* and *Gentiana*.

There are a number of species of plants, the aerial parts of which do not wither away at the end of the growing season but persist as hard, much branched cushions and polsters, though lacking underground persistent stems, rootstock, etc. Such plants are covered over by the winter snow and lie totally buried, without injury, inside the snow throughout the whole winter and early spring. When the snow has melted next summer, their buds reawaken and unfold new foliage and the plant as a whole resumes its growth, producing more and more branches, year after year, to give rise to hemispherical or dome-shaped growths. These include not only the shrubs already mentioned, but also some interesting typical herbs and shrub-like herbs. It is not often easy in the field to classify a plant as a typical herb or shrub, particularly in the case of the mat and cushion plants. Many plants that grow as shrubs at elevations of about 3000 m however spread out into low and matted growths at elevations above 3900 m, where they can hardly be described as shrubs. The structure and general appearance of nearly all the plants vary within wide limits, depending on differences in altitude and the general severity of the conditions prevailing in the locality.

In a number of plants the internodes of the aerial stems are greatly shortened or even totally suppressed, so that the leaves are largely or wholly all radical; the cauline leaves. if any, are generally very small and few. This condition is commonly observed in species of *Caltha*, *Trollius*, *Callianthemum*, *Braya*, *Oxalis*, *Inula*, *Cremanthodium*, *Senecio*, *Gerbera*, *Hieracium*, *Taraxacum*, *Lactuca*, often also in *Scorzonera*, *Primula* and largely in *Lagotis* when found growing near the upper limits of its altitudinal range. In such extreme cases the leaves are greatly crowded together as, for example, in *Acantholimon*, or whorled as in *Pedicularis* or imbricate and even rosulate as in *Alchemilla*, *Sempervivum*, *Sedum*, *Androsace* (Pl. XVIII), *Thylacospermum*, *Cassiope*, *Morina* and *Diapensia*.

The shortening and suppression of internodes and the formation of rosettes are common characters of the alpine-zone plants on nearly all high mountains, in the subarctic, arctic and sub-antarctic areas. The underlying factors of the persistent rosette formation in these different ecosystems are complex and naturally different. Rosettes are no doubt readily explained by very low growth rate, resulting in shortening of the internodes. Intensity of sunlight perhaps inhibits the elongation of stem at high altitude—the long summer daytime of the high altitude and the "midnight" sun of the high latitudes may be perhaps important factors in suppressing internodes. The temperature regime and retardation of the flow of sap over long distances of internodes, because of excessive atmospheric cold and snow, must not also be overlooked. It may perhaps be safely assumed that at least at high altitudes the suppression of internodes is a response to changes of atmospheric temperature and effects of intense light. Rosette formation, though common to the high altitude and high latitudes, may actually be very different responses. All rosettes at high altitude on the Himalaya close tightly into compact buds, resembling a miniature cabbage, at night and reopen only after the sunrise. The central growing point, within the rosette, remains always closely and tightly ensheathed by the young leaves, which also continue to remain folded until the scape pushes upwards as it grows. The closing of rosettes at night is apparently related to fall in air temperature.

Many species are remarkably quite devoid of an aerial stem at extreme high elevations. This is, for example, typically the case in species of *Ranunculus*, *Oxygraphis*, *Callianthemum*, *Meconopsis*, *Parrya*, *Cochlearia*, *Eutrema*, *Inula*, *Saussurea*, *Jurinea* and *Microula*. The absence of an aerial stem and the fact that all leaves are radical are conditions correlated with the blade being narrowed at base, broader apically and showing a strong tendency to be spatulate, obovate or oblanceolate and pinnatifid or pinnatisect as in Cruciferae, many Compositae and others. The generally great physiological age of foliage formation probably also would explain, at least partly, the predominance of dentate, lobed and pinnatisect conditions of the leaves at high elevations.

The virtual elimination of internodes and of the aerial stems extends even to the inflorescence axis. In a great variety of plants like *Anemone*, *Ranunculus*, *Oxygraphis*, *Caltha*, *Callianthemum*, *Draba*, *Acantholimon*, *Primula*, etc., the inflorescence is scapigerous. The scapes are nearly always leafless and perhaps represent the only erect part of an otherwise low-growing plant. The peduncles and pedicels may also be conspicuously shortened and many plants like *Eutrema*, *Cochlearia*, *Alyssum*, etc., have their flowers in corymbs or in shortly peduncled heads as in *Oxytropis*, *Trifolium*, etc. The same tendency should explain the dominance of compound umbels of Umbelliferae, compound heads of the Compositae being gathered together is semi-umbellate or corymbose clusters. Massed spikes, verticillate bunches, short and compound umbels, and corymbose heads are the more common and dominant types of inflorescence at higher elevations. Single flowers and loose panicles are mostly found in more or less sheltered localities, where the general conditions do not appear to be so extreme. The same factors and rapid and massive flow of sap to numerous flowers, all of which open about the same time, favour the shortening of the main inflorescence axis, peduncles, etc. The atmospheric cold should be expected to retard the flow of sap over elongated shoot axis, so that its abbreviation would counteract the effects of cold and favour adequate and rapid supply of nutrients. The massing and clustering of flowers at the tip of a generally leafless scape are perhaps parts of the same specializations and must be closely correlated with the formation of foliage rosettes. The rapid development of the inflorescence may perhaps inhibit the normal elongation of internodes and in this way indirectly favour the rosette formation. At any rate, the leaf rosettes, the long leafless scapes and the massing of flowers seem to be all interlinked characters, apparently related to the conditions prevailing at ground level in the high altitude ecosystem.

The foliage of high altitude plants presents a number of interesting specializations. There is nearly every gradation from the most delicate to the rather heavily coriaceous foliage, depending mainly on the special habitat. The preponderance of small-sized and coriaceous leaves is widely believed to be the result of action of ultra-violet. Dense woolliness is also known to be produced in response to high intensity of ultra-violet. Some of the plants with rather conspicuously thick leaves, particularly in the higher elevations, belong to *Sedum*, *Sempervivum* and certain species of *Androsace*. Nearly all parts of numerous plants, but particularly the leaves, are hoary, densely pubescent, villous, woolly or even cottony. For example, *Cheiranthus* is typically hoary and *Draba* is often tomentose. Among the pronouncedly woolly or silky haired plants are *Cimifuga*, *Meconopsis*, *Alyssum*, *Sisym-*

brium, *Potentilla*, *Aster*, *Erigeron*, *Leontopodium*, *Tanacetum*, *Eritrichium*, etc. Profusely cottony plants include *Anaphalis*, *Helichrysum* and *Saussurea* among others. On account of the dense cottony hairs, *Saussurea gossypiphora* is used in Lahaul-Spiti as tinder by nomadic shepherds. Woolly plants are characteristic of sub-arctic and sub-antarctic ecosystems also, but the varying underlying causes of woolliness evoke responses and secondary effects. Woolliness is associated with insulation and retardation of evaporation; the entrapped air in between the long hairs has great significance in both these effects. Effective protection is also provided by dense cover of dried leaf bases, expanded petioles, tufted branches, scaly persistent parts, etc., that cover buds and perennial parts.

The high altitude ecosystem of the Northwest Himalaya is remarkable for the dominance of yellow or pink to red coloured flowers; white flowers are not also infrequent. Only exceptionally are the flowers violet or blue and indeed hardly over a dozen genera of plants have blue coloured flowers, particularly at extreme elevations, viz., *Aquilegia*, *Delphinium*, *Aconitum*, *Meconopsis*, *Saussurea*, *Exacum*, *Gentiana*, *Jaeshkea*, *Pleurogyne*, *Lindelofia*, *Paracaryum*, *Eritrichium*, *Rochelia*, *Lagotis* and *Hyssopus*. On the other hand the plants with yellow coloured flowers are met with in over seventy genera like *Anemone*, *Ranunculus*, *Oxygraphis*, *Caltha*, *Trollius*, *Adonis*, *Aconitum*, *Cimifuga*, *Stylophorum*, *Hypecoum*, *Corydalis*, *Cheiranthus*, *Barbarea*, *Alyssum*, *Draba*, *Cochlearia*, *Sisymbrium*, *Christolea*, *Chorispora*, *Erysimum*, *Silene*, *Biebersteinia*, *Thermopsis*, *Melilotus*, *Caragana*, *Oxytropis*, *Geum*, *Potentilla*, *Saxifraga*, *Parnassia*, *Sedum*, *Bupleurum*, *Hieracium*, *Morina*, *Aster*, *Leontopodium*, *Anaphalis*, *Inula*, *Tanacetum*, *Cremanthodium*, *Doronicum*, *Werneria*, *Cousinia*, *Carduus*, *Gerbera*, *Crepis*, *Taraxacum*, *Lactuca*, *Scorzonera*, *Centaurea*, *Primula*, *Arnebia*, *Ononis*, *Scrophularia*, *Pedicularis*, *Nepeta*, *Phlomis*, *Stellera*, *Lloydia*, *Gagea*, etc. Nearly sixty genera have species with reddish or pink coloured or rose coloured or purple flowers. The common examples of these belong to *Anemone*, *Adonis*, *Aconitum*, *Hypecoum*, *Corydalis*, *Cheiranthus*, *Cochlearia*, etc. The flowers of *Sisymbrium*, *Braya*, *Parrya*, *Chorispora*, *Cardamine*, *Viola*, *Gypsophila*, *Thylacospermum*, *Geranium*, *Ononis*, *Trifolium*, *Parochaetus*, *Oxytropis*, *Saxifraga*, *Sedum*, *Epilobium*, *Trachydium*, *Pimpinella*, *Pleurospermum*, *Valeriana*, *Triplostegia*, *Morina*, *Scabiosa*, some forms of *Aster*, *Ergieron*, *Cousinia*, *Carduus*, *Saussurea*, *Jurinea*, *Gerbera*, *Cassiope*, *Acantholimon*, *Diapensia*, several species of *Primula*, *Androsace*, *Swertia*, *Lindelofia*, *Myosotis*, *Macrotomia*, *Onosma*, *Veronica*, *Pedicularis*, *Lagotis*, *Origanum*, *Thymus*, *Hyssopus*, *Calamintha*, *Dracocephalum*, *Polygonum* and some species of *Allium* are either rose, pink or purple or deep red coloured. The plants with white flowers belong to some forms of *Anemone* and *Caltha*, *Isopyrum*, *Delphinium*, *Aconitum*, *Callianthemum*, some examples of *Parrya*, *Arabis*, *Alyssum*, *Draba*, *Cochlearia*, *Sisymbrium*, *Capsella*, *Braya*, *Crambe*, *Carda-*

mine, *Thlaspi*, *Eutrema* and *Iberidella*. Other white flowered plants are species of *Lychnis*, *Cerastium* and some *Trifolium* and one species of *Melilotus*, *Spiraea*, *Geum*, *Fragaria*, *Saxifraga*, *Parnassia*, *Trachydium*, *Hieracium*, *Valeriana*, *Aster*, *Allardia*, *Artemisia*, *Cousinia*, *Carduus*, *Gerbera*, *Cassiope*, *Exacum*, *Myosotis*, *Rochelia*, *Arnebia*, *Eremurus*, *Allium*, etc.

The general appearance of the high altitude vegetation of the Northwest Himalaya, when viewed from a distance, is remarkably striking and strongly reminds one of the Pamirs, rather than of the rest of the Himalaya or the Alps. While the general colour of the vegetation of the tundra and arctic north and even also of the Eastern Himalaya is known to tend towards blue-green rather than green, the Northwest Himalayan high altitude vegetation is predominantly yellowish-green or has even an excess of brown. These differences have been explained as part of the specialization for effective reflection of up to 55% of the infra-red of the solar radiation in the case of blue-green vegetation. The yellowish-green or brown vegetation reflects much less than 16% of the infra-red of the sunrays [94]. These peculiarities become meaningful if we recollect that there is an excess of blue and ultra-violet in the high altitude environment. The recent studies of Popova [94] have shown that the pigments of the high altitude plants do not differ basically from those of the lowland plants, but the chlorophyll content is significantly low in the high altitude species. Bukharin [14] has made the interesting observation that carotene content is also generally low in the vegetation of the Pamirs. Like in the Pamirs, the high altitude vegetation of the Northwest Himalaya, and particularly on the inner ranges, is characteristic for its brown appearance, which is quite conspicuous from even a distance.

5. ECOLOGICAL TYPES AND AFFINITIES

The plants of the high altitude ecosystem belong to diverse communities and form characteristic associations, depending upon the habitats. Some of the dominant types may be considered as examples.

1. Petrophile communities of barren rock surface (Pl. VII, VIII) and among large boulders, often exposed as nunatak above the snowline and sporadically moistened by thin films of melt-water. The character species belong generally to many eurytype lichens like *Placodium*, *Parmelia*, *Xanthoria*, *Rinodina*, *Aspicilia*, *Ochrolechia*, *Peltigera*, *Cladonia*, *Lecanora*, *Lecidea*, *Umbilicaria*, *Gyrophora*, *Rhizocarpon*, *Stereocaulon*, *Dactylina*, *Cetraria*, etc. In addition to xerophiles, there are also a number of petrophile mesophytes like occasional *Saxifraga*, mosses, etc. Petrophile hygrophiles include the liverworts, mosses and specially eurytypes like *Aulacomnium*, *Bryum*, *Drepanocladus*, *Climacium*, *Campylium*, *Rhytidium*, *Polytrichium*, *Pseudostereodon*, *Thuidium*, *Abietinella*, *Eucalypta*, *Oncho-*

phorus, *Myurella*, *Tortella*, *Grimmia*, *Orthotrichium*, etc. It is interesting to observe that some of these genera and even species are found at comparable habitats on the Pamirs and the Central Tien Shan [43, 44, 45, 46, 69].

2. Snow edge hygrophile communities of delicate, aerial-annuals, with perennial underground parts, generally herbaceous, typically confined to the margin of the melt-water streams, edge of the melting winter snow and other humid localities. The character species include, in addition to moss, members of genera like *Caltha*, *Trollius*, some *Primula*, etc. (Pl. I, 2).

3. Snow surface communities of peculiar Algae and Volvocales like *Chlamydomonas*, *Scotiella*, *Trochiscia*, *Mycanthococcus*, *Romeria*, *Pleurococcus*, *Sorochlora*, *Raphidonema*, *Siderocarpus*, *Leptothrix*, etc., that grow almost exclusively on the surface of snow fields and frequently impart a striking yellow or greenish-yellow colouration to the snow.

4. Glacier zone communities of cryophytes, with the character species, belonging generally to genera like *Primula*, *Waldheimia*, *Pachypleurum*, sometimes *Oxytropis*, some *Potentilla*, *Poa alpina*, *Saxifraga hirculus*, *Draba*, *Saussurea sorocephala*, *Parrya excapa*, *Tanacetum*, occasionally *Nepeta*, etc. In glacial moraines may often be found *Ranunculus*, *Callianthemum*, *Oxygraphis*, *Sedum*, *Oxyria*, *Draba*, *Saussurea*, *Saxifraga*, *Chorispora excapa*, *Thylacospermum caespitosum*, etc.

5. The grassy meadow (Pl. IX, X, XII, XV) communities, constituting the mountain frigid steppes, dominated by mesophytes and even by many hygrophiles like *Gentiana*, *Scabiosa*, *Leontopodium*, *Polygonum*, some *Astragalus*, *Pachypleurum*, *Veronica*, *Aster alpinus*, *Saussurea*, *Gypsophila*, *Primula*, *Anemone*, *Delphinium*, *Galium*, *Pedicularis*, *Androsace*, *Oxytropis*, *Eremurus*, *Hedysarum*, etc., at lower elevations immediately above the timberline. The vegetation of meadows may also comprise species of *Phlomis*, *Sanguisorbus*, *Scabiosa*, *Pachypleurum*, *Aconitum*, *Ranunculus*, *Fritillaria*, *Allium*, *Corydalis*, *Doronicum*, *Artemisia*, *Trollius*, *Aquilegia*, *Valeriana*, *Lindelofia*, *Adonis*, *Eremurus*, *Poa*, *Festuca*, *Agropyrum*, *Bromus*, *Carex*, *Cobresia*, etc.

At higher elevations the meadows are generally covered by growths of *Poa alpina*, *Festuca*, *Polygonum viviparum*, *Oxygraphis glacialis*, *Gentiana*, *Primula*, *Ranunculus*, *Allium monadelphum*, *Erigeron*, *Pedicularis amoena*, *Oxytropis immersa*, *Pachypleurum alpinum*, *Draba stenocarpa*, *Inula rhizocephala* (Fig. 27), *Astragalus*, *Aster alpinus*, *Cirsium*, *Myosotis*, *Lagotis*, *Potentilla*, *Dracocephalum*, *Saxifraga hirculus*, *Androsace*, *Callianthemum*, etc. Some plants that are typical of the lower elevation grassy meadows, but often found higher, include species of *Carex*, especially *Carex stenocarpa*, *Cobresia*, etc. At higher elevations the *Cobresia*-meadow is a characteristic feature of the Northwest Himalaya; we find here several species of *Cobresia* like *C. royleana*, *C. bellardii* and other plants like *Draba*, *Swertia marginata*, *Pachypelurum*, *Saxifraga hirculus*, *Pleurogyne*, etc. At elevations of 4200–4500 m, particularly on the north slopes, this formation includes *Carex*, *Primula*, *Artemisia* with *Cobresia stenocarpa*. Most of these plants are typically mesophytes, but there are a number of xerophytes also. Although all the formations and plant associations have not been fully studied so far, mention may be made of the following important ones: 1) *Phlomis*-formation, 2) *Geranium*-formation, 3) *Dracocephalum diversifolium*-formation, 4) *Iris brevituba*-formation, 5) *Festuca*-formation, 6) *Polygonum*-formation, etc. The *Phlomis*-formation is composed of species of *Phlomis*, *Geranium*, *Senecio*, *Stipa*, *Polygonum*, *Caragana*, *Cobresia*, *Leontopodium*, *Festuca*, etc. The *Geranium*-formation has the associated species of *Alchemilla*, *Bromus*, *Phlomis*, *Scabiosa*, *Poa*, etc., besides *Polygonum* and *Iris*. Some other formations of the high alpine zone grassy meadows observed so far include *Alchemilla*-formation, *Ligularia*-formation, *Allium*-formation, *Poa alpina*-formation, *Polygonum viviparum*-formation, *Leontopodium*-formation, *Taraxacum*-formation, etc. The cryophyte meadow formations comprise those of *Acantholimon*, *Oxytropis*, *Saussurea*, *Potentilla*, etc., each composed of typical character species. The plant formations and associations at extreme elevations need much extended observations and are largely unknown at present; the ones mentioned here merely serve to illustrate the phenomena.

When we survey the high altitude vegetation of the Northwest Himalaya

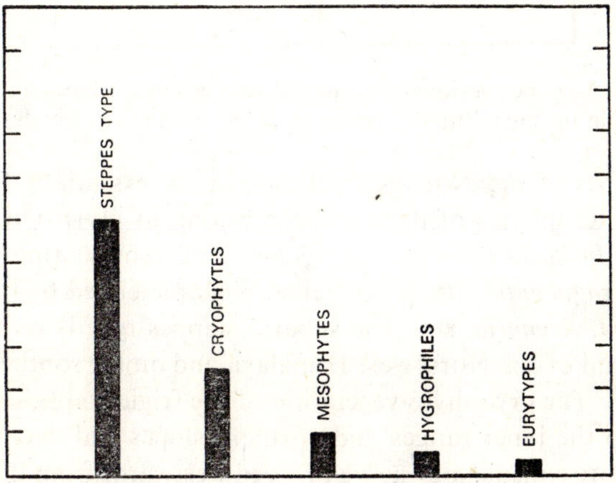

Fig.13. The relative abundance of differnet ecological types of plants in the high altitude ecosystem of the Northwest Himalaya.

in its entirety, we find that the following ecological types are generally dominant: 1) subarid types, including the mountain steppes, mountain savannah [69], mountain cryophyte-xerophyte; 2) humid types on the subalpine and alpine grassy meadows; 3) cryophytes of cushion plants; 4) petrophytes and 5) eurytypes. The relative abundance of the different ecological types, summarized in Fig. 13, shows the pronounced dominance of the steppes type and cryophytes over the mesophytes; hygrophiles and eurytypes are minor components in the high altitude ecosystem of the Northwest Himalaya. The abundance of the different ecological types at different altitudes shows a marked difference; the steppes and cryophytes are greatly dominant at higher elevations, the mesophytes are relatively scarce at elevations higher than 5000 m (Fig. 14).

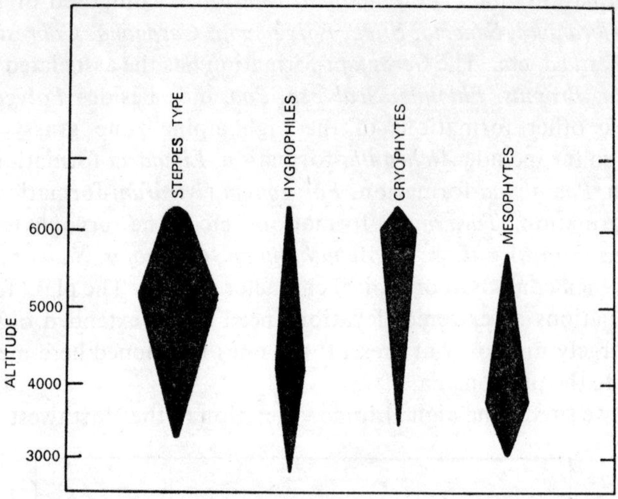

Fig. 14. Zones of dominance of different ecological types of plants in the high altitude ecosystem of the Northwest Himalaya.

The steppes of the Northwest Himalaya are essentially frigid steppes. The typical xerophytes of these steppes belong to *Oxytropis*, *Onobrychis*, *Scorzonera*, *Jurinea*, *Lagotis*, *Astragalus*, etc., representing what may be best called *Tragacantha*-steppes, often also characterized by the presence of *Acantholimon*, *Arenaria*, etc. The subarid steppes type is rather marked in the eastern end of the Northwest Himalaya and on the southern slopes and outer ranges. The cryophyte vegetation of the frigid steppes is particularly dominant on the inner ranges and northern slopes and becomes also progressively more conspicuous as we proceed westwards from the defile of the R. Sutlej. In the north it merges, without an abrupt break, into the frigid steppes of the Pamirs.

In its ecological characters, the high altitude vegetation of the Northwest Himalaya differs very markedly from that of the Kumaon or Central Himalaya; its affinities are very clearly with the vegetation of the extreme high elevations, in the more arid interior of the Kumaon and Nepal areas, rather close to Tibet border. The ecological characters reveal an interesting transition from the east to the west and also from the south to the north, from the Pir Panjal northwards across the parallel ranges to Ladakh and finally merging, imperceptibly in the high Pamirs. As we move along westwards, above the timberline, on the Himalaya, from Assam in the east, there is a gradual but nevertheless clearly recognizable gain in the general dominance of the steppes character, with corresponding decline in the abundance of hygrophiles and mesophytes (Fig. 15).

From the ecological point of view, as well as from the biogeographical standpoint, discussed further below, the Northwest Himalaya must, most appropriately, be described, therefore, as the extreme southern border of the Pamirs-Central Tien Shan ecosystem complex. This conclusion is supported by far-reaching similarities of the formations and sequence of ecological types, besides other features already discussed. In many cases, there is very close identity between the Northwest Himalaya and the Pamirs. Korovin [69] has, for example, listed the plants characteristic of the frigid steppes of the Central Tien Shan as consisting of *Gentiana aurea*, *G. olgae*, *G. kirilovii*, *G. prostrata*, *Scabiosa alpestris*, *Leontopodium alpinum*, *Melandrium apetalum*, *Polygonum bistorta*, *P. alpinum*, *Astragalus alpinus*, *Pachypleurum mucronatum*, *Veronica apicata*, *V. spuria*, *Aster alpinus*, *Saussurea salicifolia*, *Gypsophila cephalotes*, *Primula algida*, *Anemone prostrata*, *Galium boreale*, *Delphinium dasyanthum*, *Pulsatilla campanella*, *Pedicularis amoena*, *P. dolichorrhiza* and *Galium verum* in the subalpine zone. The similarity of the composition of this type in the Northwest Himalaya, listed above is striking. In the Alai-Pamirs, the affinity is even better demonstrated by *Androsace darvasica*, *Oxytropis darvasica*, *Polygonum bucharicum*, *Eremurus korshinskyi*, *Astragalus nigrocalyx*, *Hedysarum flavescens*, etc. The mountain meadows, in the lower alpine zone elevations, of the Pamirs and the Central Tien Shan, have vegetation typically composed, as on the Northwest Himalaya, of many identical or closely related plants like *Phlomis oreophylla*, *P. pratensis*, *P. canescens*, *Sanguisorbus alpestris*, *Scabiosa alpestris*, *Pachypleurum mucornatum*, *Aconitum rotundifolium*, *A. excelsum*, *A. soongoricum*, *Delphinium confusum*, *D. elatum*, *D. poltoratzkii*, *Ranunculus grandifolius*, *Fritillaria pallidiflora*, *Gentiana decumbens*, *G. tianschanica*, *G. kaufmanniana*, *Pedicularis macrochila*, *Allium platyspathum*, *A. polyphyllum*, *A. monadelphum*, *Corydalis gortschakovii*, *Doronicum altaicum*, *Artemisia aschurbajevii*, *A. frigida*, *A. lehmanniana*, *Trollius dshungaricus*, *T. altaicus*, *T. asiaticus*, *Aquilegia lactiflora*, *Valeriana tianschanica*, *Lindelofia*, *stylosa*, *Adonis chrysocanthus*, *Eremurus altaicus*, *Poa relaxa*, *Festuca rubra*, *Bromus turkestanicus*, *Carex stenocarpa*, *Cobresia schoenoides*, etc. The similarity of the alpine meadow vegetation is revealed by the presence in the Pamirs of *Poa alpina*, *Festuca kryloviana*, *F. kirilowii*, *F. alatavica*, *F. tianschanica*, *Poa litwinowiana*, etc., species which are closely related to those of the Northwest Himalaya. The other alpine meadow plants listed by Korovin (*op. cit.*) from the Pamirs, which again show the pronounced affinity to the

Northwest Himalaya, include *Polygonum viviparum*, *Oxygraphis glacialis*, *Gentiana algida*, *G. nutans*, *G. kaufmanniana*, *G. grombtschevskii*, *Primula nivalis*, *P. algida*, *Ranunculus alberti*, *R. rufosepalus*, *R. olgae*, *R. rubrocalyx*, *Allium monadelphum*, *Erigeron auranticus*, *E. azureus*, *E. oreodes*, *E. lachnocephalus*, *Pedicularis violascens*, *P. dolichorrhiza*, *P. amoena*, *P. dubia*, *Oxytropis immersa*, *Pachypleurum alpinum*, *Draba oreades*, *D. altaica*, *D. stenocarpa*, *D. melanopus*, *D. talasica*, *Inula rhizocephala*, *Astragalus tibetanus*, *Astragalus alpinus*, *Aster alpinus*, *Lagotis glauca*, *Potentilla gelida*, *P. hololeuca*, *Dracocephalum discolor*, *Saxifraga hirculus*, *Androsace lehmanniana* and *Callianthemum alatavicum*. Some of the local associations and formations like the *Cobresia*-formation, *Phlomis*-formation and the *Gentiana*-formation, mentioned earlier, are fully matched by parallel species and habitats in the Pamirs and Central Tien Shan.

Ecologically the high altitude region of the Northwest Himalaya is a part of the subarid Middle Asiatic steppes, modified secondarily in small pockets into frigid steppes, composed of cryophytes at sufficiently high elevations. The typical features of mountain steppes, particularly of Middle Asia, are described in detail by Korovin [69].

6. THE ECOLOGICAL TRANSITIONS

We may now present briefly the general ecological transitions of the Himalaya as a whole, from its eastern end to the extreme western. Fig. 15 is a simplified diagrammatic representation of this transition; for a really accurate picture we need, however, a three dimensional sketch. In the Eastern Himalaya the high altitude vegetation exhibits many characters, which are highly reminiscent of tropical high mountains. There is a great preponderance of hygrophiles and mesophytes are by no means few, but steppes forms and xerophytes are sparse and are also confined largely to extreme elevations and northern slopes of the mountain ranges. The picture remains almost the same in the eastern sector of the Central Himalaya, but in the western sector, we observe the increasing dominance of mesophytes and the hygrophiles become at the same time less and less dominant. The steppes and xerophytes are distinctly perceptible elements. The Western Himalaya foreshadows the ecology of the Northwest Himalaya, and must be considered as a connecting link between the Eastern and Northwest Himalaya. The steppes and xerophile types become common enough, though the mesophytes are still the dominant types here. Hygrophiles may be considered as almost evanescent. The Northwest Himalaya is dominated by arid and frigid steppes and though mesophytes are minor elements, the hygrophiles are on the whole extremely sparse and confined to small and isolated pockets of favourable localities. The Northwest Himalaya merges in the west and in the north with the steppes of the Hindu Kush and the Pamirs.

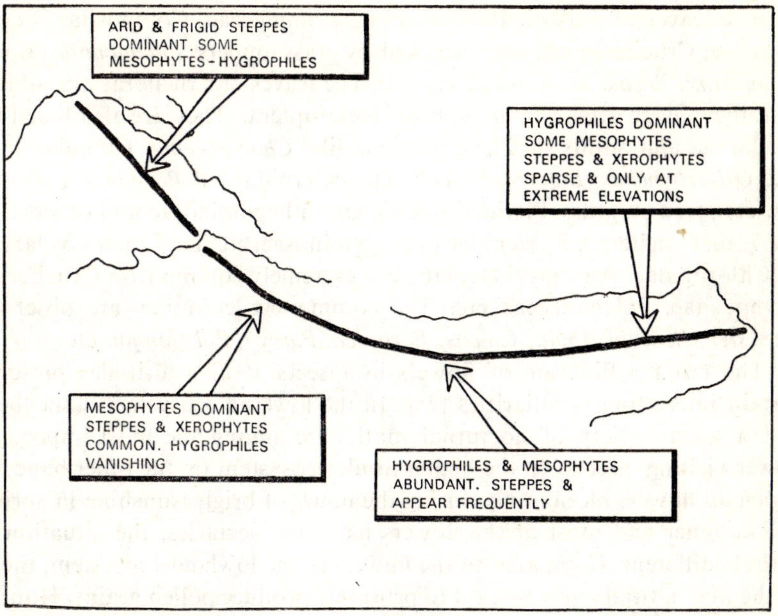

Fig. 15. The east-west ecological succession in the Himalaya, showing the pronounced fundamental difference between the Northwest Himalaya and the eastern end of the Himalaya.

The high altitude vegetation of the Northwest Himalaya is essentially an arid steppes vegetation, which has been secondarily modified as frigid steppes by the uplift of the mountains.

7. INTERRELATIONS BETWEEN HIGH ALTITUDE PLANTS AND INSECTS

Any discussion of the general ecology of high altitude plants would be incomplete without some reference to the interrelations with insects. The high altitude insects and their interrelations to the plants of the Himalaya and Pamirs have been outlined in an earlier work [76, 77]. We may consider here, for the sake of completeness, two principal associations of plants and insects at high altitudes.

The most important and perhaps also the primary plant-insect association at high altitude is phytophagy. Although phytophagous insects are by no means dominant at high altitudes [75, 76], the forms that occur are generally associated more or less obligatorily with certain families, genera or even species of plants. The larvae and adults of these insects either devour the leaves and flowers or mine in leaves or tunnel inside the rootstock, stems, fruits, etc. Leaf-miners are apparently progressively common at higher elevations.

The leaves of grasses, Polygonaceae, Primulaceae, Compositae, Leguminosae, Cruciferae, etc., are attacked by grasshoppers *Sphingonotus*, *Gomphomastax*, *Bryodema*, *Conophema*, etc. The leaves of Cruciferae are subject to rather severe attack by nymphs of Heteroptera. They are also holed by the larvae and adults of Chrysomelidae like *Chaetocnema*, Curculionidae like *Otiorrhynchus*, *Blosyrodes*, etc. The caterpillars of *Parnassius*, *Argynnis*, etc., feed on grass, Pieridae like *Colias* on Leguminosae and others like *Pieris* on Cruciferae, Lycaenidae on Leguminosae, etc. Leaf-mines by larvae of Chloropidae, Agromyzidae, etc., are extremely common on Cruciferae, Compositae, Polygonaceae, etc. The commonest leaf-mines are observed on *Aster*, *Senecio*, *Inula*, *Cnicus*, *Barbarea*, *Parrya*, *Polygonum*, etc.

The cross pollination of flowers by insects at high altitudes presents certain interesting peculiarities [75]. In the lowland ecosystem, honeybees and a great variety of nocturnal moths are among the most important flower visiting insects. In the high altitude ecosystem on the other hand, in which all flowers bloom only during the hours of bright sunshine in spring and summer and most of the flowers lack also nectaries, the situation is entirely different. Compared to the flowers in the lowland ecosystem, those of the high altitude may be said to profusely produce pollen grains. Honeybees are also absent at extreme elevations and in their place we find many species of the pollen-gathering *Bombus*, (Pl. XXIV) which are all diurnal. Butterflies like *Argynnis*, *Colias*, *Pieris*, *Parnassius*, *Papilio*, etc., are among the important pollinators of the flowers of Cruciferae, Leguminosae, Compositae, Umbelliferae, *Potentilla*, etc. Next in importance as flower visiting insects are Diptera like Anthomyiidae, Syrphidae, etc., which visit flowers of Umbelliferae, *Primula*, *Androsace*, etc. (Pl. XIX). The subject of cross pollination of the flowers by insects in the high altitude ecosystem is urgently in need of fuller investigation than has been undertaken in the past.

8. SUMMARY

This chapter presents a broad outline of the synecology of high altitude plants and describes the salient characters of the high altitude vegetation.

The reactions of plants to the environmental conditions, prevailing at high altitudes on mountains, differ basically from those of other organisms. Atmospheric cold and the increasing deficiency of oxygen in the air at extreme high elevations do not by themselves exclude plant life. The great importance of carbon dioxide deficiency in the high altitude air limits photosynthesis and has therefore greater ecological importance than the oxygen deficiency or cold. The limiting factors for plants at extreme high elevations depend primarily on the phenomena of melting of snow and ice into thin films of moisture by the intense daytime insolation in summer, during the hours of bright sunshine. Owing to these peculiarities, plants occur at much higher elevations than most animals, except perhaps some insects; many plants flourish at extreme high elevations, above 6300 m.

Then follows a brief account of the distinctive features of high altitude vegetation. Particular attention is drawn to its marked difference from the subarctic and arctic vegetations. The xerophytic characters of the high altitude vegetation cannot be equated with those of the tundra and arctic cold deserts or even with those of hot deserts, which are all completely lowland ecosystems.

The abbreviation of the growing season and its effects on high altitude plants are described.

In a brief account of the general physiognomy of the high altitude vegetation of the Northwest Himalaya, particular emphasis is laid on the characters that serve to distinguish the high altitude plants from those of the lowland ecosystems in cold and subarid regions. A pronounced shortening of internodes, formation of rosettes, suppression of aerial stems, dwarfing, tufted growth, differentiation into aerial annual and underground perennial growths, dominance of woolly and cottony outgrowths, predominance of yellow, pink or red coloured flowers, massing and clustering of flowers and the compounding of inflorescence, etc., are considered as parts of high altitude specializations in plants.

The major plant communities and formations are outlined, with suitable examples. The ecological types and their abundance and altitudinal distribution are discussed. The ecological affinity of the Northwest Himalaya to the Alai-Pamirs is pointed out.

Considered in its entirety, the high altitude vegetation of the Northwest Himalaya is dominated by its arid or subarid and frigid steppes character; mesophytes are not abundant and hygrophiles are extremely sparse.

The ecological east-west transition in the Himalaya is outlined. In the eastern end, the Himalaya is dominated by hygrophiles and mesophytes and steppes forms are extremely sparse.

As no account of the general ecology of plants is complete without some reference to their interrelations with insects, a brief mention is made of the principal cross-pollinators of high altitude flowers; most flowers are Diptera flowers or Lepidoptera flowers and only few are Hymenoptera [*Bombus*] flowers.

Chapter IV

Some Typical High Altitude Plants

1. INTRODUCTION

THIS chapter deals with some of the common high altitude plants observed by us in the Northwest Himalaya. The typical high altitude plants of the Northwest Himalaya comprise about three or four hundred species, belonging to some two hundred genera, distributed under forty families. The dominant families are Ranunculaceae, Cruciferae, Caryophyllaceae, Compositae, Boraginaceae, Umbelliferae, Labiatae and Gramineae. Geraniaceae, Leguminosae, Scrophulariaceae, Primulaceae, Cyperceae, Saxifragaceae and Gentianaceae come next in importance. Some of the exclusively mountain groups like Fumariaceae, Dipsacaceae, Valerianaceae, Primulaceae, etc., occupy, as may be expected, an important place in the high altitude ecosystem of the region. The frequency of different families is shown in Fig. 16. Certain families common enough in the flora of temperate lands, such as for example, Malvaceae, Convolvulaceae, Rubiaceae, Acanthaceae, Verbenaceae, Vitaceae, Rhamnaceae, etc., and some others generally considered as semi-altitude plants (Hypericaceae), are conspicuous by their total absence at high altitudes in the Northwest Himalaya. The absence of the climbing Convolvulaceae and Vitaceae in the high altitude ecosystem of the region seems understandable, but the absence of xerophytic and mesophytic forms of Acanthaceae, even in the semi-arid and arid localities, is remarkable. It must be pointed that these families are well represented even in the upper montane forest zones of the Northwest Himalaya, and particularly on the southern slopes of the mountain ranges, but they disappear, with perhaps the exception of some Rhamnaceae, far below the subalpine zone. Although we cannot describe any single family as an exclusively high altitude group, we may consider the families listed here as more or less typical of the high altitude ecosystem of the Himalaya.

The plants listed here are merely illustrative and not exhaustive. No doubt many species, which occur at high altitudes, may have been omitted; we have not come across them and future explorations are bound to bring to light many more interesting species.

During our explorations, we have faced one constant difficulty. Most of the high altitude plants are extremely variable, particularly at extreme high elevations, near the snowline, so that the individuals of the same species, collected from different altitudes, present markedly different appearances. It has not, therefore, always been an easy task to place many plants in their proper species. Many species are also often separated from other closely allied ones by extremely trivial differences, easily overlooked. It is almost impossible in such cases to effect a specific placement unless we have recourse to authoritatively named herbarium specimens for comparison. An

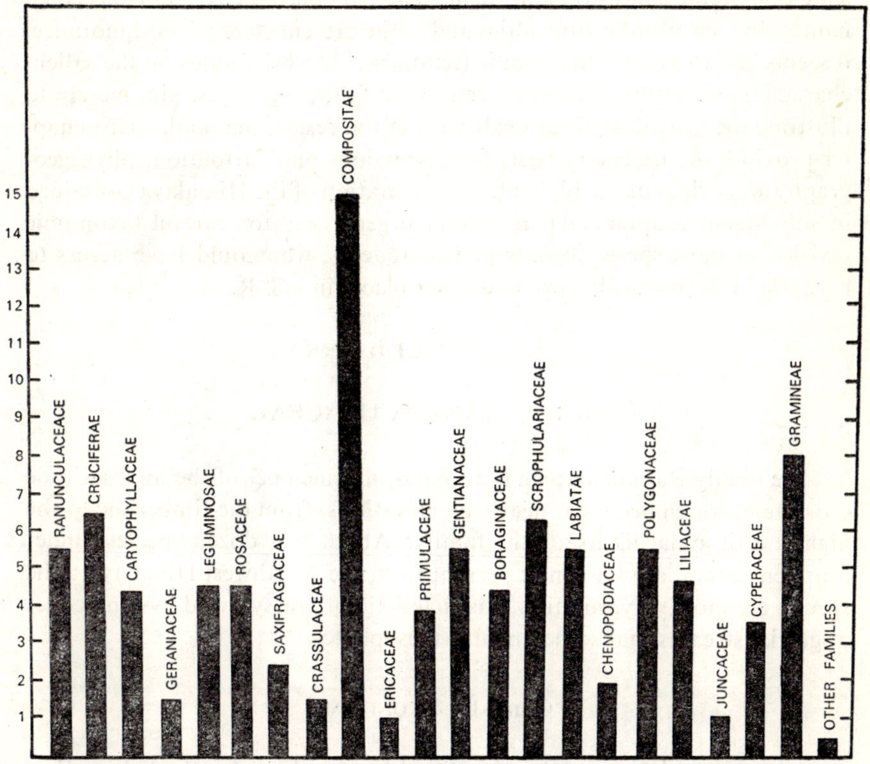

Fig. 16. Histograms of the mean frequency distribution of the major families of Phanerogams in the high altitude ecosystem of the Northwest Himalaya, to show the general abundance of different families. They merely serve to indicate the broad tendencies, as estimated from available data and allowances must be made for minor fluctuations.

accurate identification is also rendered difficult by the fact that fresh plants often present a very different appearance from herbarium material. Besides, a number of species, described from the Himalaya, are perhaps no more than local variants of and are likely to prove identical with those already known from the Pamirs, Central Tien Shan and the Altai Mountains. It seems unfortunate that while the taxonomic identification of the high altitude plants of the Pamirs and areas farther north may be considered to have been accomplished fairly accurately, most plants from the high altitude areas of the Northwest Himalaya can only be placed tentatively in their species. In the following pages, wherever species are named, it must be clearly understood that the specimens studied by us approached closest to the species so named or at least they could not be placed more satisfactorily elsewhere. The synonymies of most species from the Himalaya are in a state of extreme confusion and we have here retained the generally accepted names. In view of these difficulties and in the present state of our ignorance, it seems best to avoid a taxonomic treatment. The brief notes on the salient characters of only the genera, given in the following pages, aim merely to illustrate the general ecology dealt with in the rest of the book. This chapter provides the necessary basis for discussions on distribution, phytogeography and origin of the high altitude vegetation of the Himalaya contained in subsequent chapters. There is very urgent need for careful taxonomic revision of most species by competent students, who could have access to herbaria in Tashkent, Frunze and other places in USSR.

2. DICOTYLEDONS

Family 1. RANUNCULACEAE

The family Ranunculaceae is an important member of the high altitude ecosystem, which occurs at nearly all elevations, from the timberline to the highest altitudinal limits of the family. About two dozen species, under fourteen genera, are known at present from the Northwest Himalaya. The species are mostly hygrophiles, characteristic of grassy meadows and edges of glacial streams, but some are also mesophytes.

Genus 1. Aconitum

Erect perennial dwarf herbs, with leaves palmately divided; radical leaves crenate, cauline leaves 2–5-partite, up to 15 cm long, but usually only about half as large. Flowers white, yellow, red or sometimes blue to purple; sepals 5, petaloid, the posterior sepal hooded; petals 2–5, the posterior 2 petals clawed; stamens many; follicles 3–5.

The genus is separated from *Delphinium*, which it resembles, in its irregular flowers, by the posterior sepal being hooded. *Aconitum heterophyllum* in bloom during April–May at elevations of 2800–3900 m and May–June at 3000–4200 m.

Genus 2. Actaea

Erect dwarf perennial herbs, not over 50 cm tall, with woody rootstock. Leaves sometimes 20–30 cm long, alternate, ternately compound; leaflets ovate-lanceolate, entire or trilobate and serrate. Flowers small, white; sepals 3–5, unequal petaloid; petals often absent, if present small, up to ten; stamens numerous; carpels single, with stigma dilated; fruit many-sided, subglobose berry.

The solitary carpel and the fruit being a berry at once identify the members of this genus. *Actaea spicata* blooms during May at 3000 m.

Genus 3. Adonis

Erect but dwarf-stemmed annuals at lower elevations, but becoming perennials at higher elevations and not exceeding 20 cm in height, with stout and rather scaly horizontal rootstock. Leaves divided; radical leaves with the petioles measuring up to 10 cm long, pinnate. Flowers yellow, sometimes also red; sepals 7–8, petaloid; petals 15–24, about twice the sepals; carpels many, with short, persistent styles; achenes planoconvex, costate, glabrous.

The presence of numerous petals separates the genus from other Ranunculaceae. *Adonis aestivalis* is perhaps the commonest species of the Himalaya, found in bloom during May–June at elevations of 3000–4000 m; the species descends into the high valleys below the timberline also.

Genus 4. Anemone
(Pl. XII)

Densely hairy or silky, tufted annual herbs, with stout, perennial and often woody rootstock, thickly clothed with old leaf-sheaths. Leaves radical, usually lobed, simply divided or also pinnatisect. Scape silky, hardly 20–30 cm tall and sometimes even decumbent; involucre leaves sessile and trilobate. Flowers about 2 cm large, usually reddish-brown, golden-yellow, white, sometimes reddish-blue and rarely purple; sepals petaloid, silky outside; petals absent; stamens numerous; carpels numerous; achenes in oval heads, strigose, elongated-cylindrical or also compressed and glabrous, with the bearded style imbedded in woolly hairs.

Anemone is readily recognized by the absence of petals, involucrate flowers and pendulous ovules. Some of the species are often placed in the genus *Pulsatilla*. The following species are mostly in bloom during May–July at elevations of 3000–5000 m: *Anemone albana* 4500 m, *A. obtusiloba* 3600–5000 m, *A. rupestris* 3000–4500 m and *A. rupicola* 3600–5000 m.

Genus 5. AQUILEGIA

Erect, much branched, dwarf perennial of highly variable appearance. Leaves compound. Flowers usually yellow, but sometimes pale purple to blue; sepals 5, petaloid; petals 5, alternating with sepals, campanulate, spreading; stamens numerous; carpels 5, sessile; fruit a follicle.

This genus is readily separated from *Caltha*, *Trollius* and *Isopyrum* by the fact that petals and sepals are equal. *Aquilegia vulgaris*, a somewhat variable plant grows in rich soil even below the timberline. At higher elevations, between 3000 and 4200 m, the plant often grows among rock and is in bloom in early August.

Genus 6. CALLIANTHEMUM

Stemless, glabrous, perennial herb. Leaves all radical, glaucous, bipinnate or decompound, with long tufted petioles, segments 2–3-sect. Scape not exceeding 10 cm long, shorter than the leaf, with a single white flower. Sepals broadly elliptic, 5, deciduous; petals 5–15, with a pit in claw, larger than the sepals; stamens numerous; carpels numerous; style short; achenes subglobose, with short style.

Callianthemum cachmericium in bloom at 2800–3900 m during May.

Genus 7. CALTHA
(Pl. VIII, XII, XXI)

Densely tufted, much branched, dwarf herb, with stout creeping rootstock; branches never exceeding 15–50 cm in height. Leaves largely radical, cordate-ovate, but sometimes also reniform, 2–3 cm large, almost entire, sometimes the older leaves may be dentate. Scape 20–25 cm long. Flowers usually yellow, but rarely also white; sepals usually 5, deciduous, petaloid; stamens and carpels numerous, the latter sessile; fruit a follicle.

The absence of petals, undivided and alternate leaves serve to distinguish this plant from others. *Caltha palustris*, with reniform or deltoid leaves, petioles long, cauline leaves sessile, leaves usually dentate, flowers yellow, in bloom during May–June at elevations 3600–4800 m. The plant grows commonly at the edge of ponds and streams and is usually associated with *Trollius*.

Genus 8. CIMIFUGA

Much branched dwarf perennial herb. Leaves 5–7 cm long, glabrous below and densely tomentose above, 2–5-divided; leaflets ovate to lanceolate, serrate. Flowers small, yellow; sepals 4–5, petaloid, deciduous; petals often modified as staminodes, 1–8 and small, clawed, bifid; stamens numerous; carpels 1–8; fruit a follicle.

This genus is recognized by its 4–8 carpels and follicle. *Cimifuga foetida* at elevations of 2500–3900 m.

Genus 9. DELPHINIUM

Much branched, spreading, generally tomentose herb, with leafy stems of not more than 30 cm length. Radical leaves 5–10 cm, trilobate or palmately 5–7-lobed or even pinnatifid. Flowers irregular, generally white, but sometimes also purple or blue, in hairy corymbs, bracts 3–5-partite; sepals 5, dorsal sepal spurred; petals 2–4, smaller than sepals, two dorsal petals spurred; stamens many; fruit about half dozen hairy follicles.

The irregular flowers, with the dorsal sepal being spurred, serve to distinguish the genus from others. *Delphinium brunonianum*, with its leaves deeply cut up to the middle and pale blue flowers, blooms in April to May or even June, on grassy meadows, near melt-water streams, at elevations from 3000 to 5500 m. According to Swan [112], this species occurs at elevations of 5500 m on Mt. Makalu.

Genus 10. ISOPYRUM

Much branched, densely tufted, dwarf annual herb, with perennial, scaly rootstock. Leaves decompound; radical leaves 2–3-divided, alternate, petiolate, with the leaflets small. Flowers white; sepals 5–6, ovate and petaloid; petals 5, clawed; stamens 10; carpels as many as 20, sessile; fruit a follicle.

The genus *Isopyrum* is differentiated from *Trollius* by its white flowers and clawed petals. *Isopyrum thalictroides* blooms during June–July at elevations of 3300–4200 m in Lahaul.

Genus 11. OXYGRAPHIS

Usually stemless, but rarely moderately tufted, dwarf-stemmed perennial herb, with fibrous roots. Leaves rounded, crenately lobed; often oblong-elliptic and entire; 20–50 mm large. Scape not exceeding 10 cm in length, with a single flower; flowers yellow; sepals 5, persistent and enlarging; petals 10–15, usually longer than sepals; stamens numerous; achenes numerous; style straight.

This genus is readily separated from *Ranunculus* by the persistent sepals and more numerous petals. *Oxygraphis glacialis*, with oblong-elliptic, usually entire or slightly crenate leaves about 2.50–3.00 cm long and petioles nearly equally long, not very common, at elevations of 3600–5500 m and *O. polypetala* at elevations of 3300–5000 m. In bloom during June.

Genus 12. RANUNCULUS

Usually stemless or sometimes dwarf-stemmed erect and tufted or decumbent perennial herbs, with rooting runners and tuberous rootstock. Radical leaves usually divided, with elongate or also abbreviated petioles; blade rounded-oblong, oblong-elliptic or often reniform, very rarely ternate; some cauline leaves palmately 3–5 partite; glabrous, sometimes somewhat pubescent but at higher elevations (above 4200 m) densely silky, coriaceous. Scape short, with panicles of most usually yellow or sometimes white flowers; sepals 3–5, caducous, sometimes hairy; petals 5–8; stamens numerous; carpels many, style short; achenes in apical heads, compressed, costate, with beak straight or hooked.

The plant is easily recognized by the herbaceous and deciduous sepals. Several species like *Ranunculus affinis* at 3000–5000 m, *R. cymbalariae* at 3300–5000 m, *R. hirtellus* at 3900–4500 m, *R. hyperboreus* at 3900–5100 m, *R. lobatus* at 3300–4900 m, *R. nivalis* at 3900–4500 m and *R. pulchellus* at 3000–5500 m are found in bloom during May and late June.

Genus 13. THALICTRUM

Erect tufted perennial herbs about 5–60 cm tall. Leaves compound, auricled. Flowers green; sepals 5, petaloid; petals absent; stamens many; carpels usually many; achenes with persistent or deciduous styles, striate, sessile.

Though many species of this genus are found about the timberline, the plant seems to be sparse at elevations above 3600 m and on rare occasions an unidentified species has been collected at an elevation of 5200 m.

Genus 14. TROLLIUS
(Pl. XXI)

Erect but dwarf perennial, hardly 15–30 cm tall, growing generally in groups, with horizontal stout rootstock. Leaves largely radical, usually 2.5–5.0 cm large, with elongate petioles, palmate, 5 or more spreading lobes, coriaceous. Flowers usually solitary; sepals 5–10 or also more, petaloid, orbicular or oblong, sometimes persistent; petals dwarfed, much smaller than sepals and hardly seen, 5–26; stamens numerous; carpels 5, sessile.

The plant is easily recognized by the small yellow petals and sessile ovaries. *Trollius acaulis*, with long petioled radical leaves, palmatifid and 5-partite, the lobes oblong-lanceolate, blooms in May–June at elevations of 3300–4200 m. Common enough in moist localities.

Family 2. PAPAVERACEAE

Genus 1. MECONOPSIS

Hairy, stemless or dwarf-stemmed, perennial herbs, with stout rootstock. Leaves 5–20 cm long, lanceolate-pinnatifid, lobed or also almost entire. Flowers generally blue or sometimes pinkish, solitary or in racemes; sepals 2, hairy, hispid or also bristly; petals 4; ovary 1-celled; capsule long.

Meconopsis aculeata is found occasionally in bloom in July at elevations of 4200 m. Rau [101] reports this plant at elevations of 3000–5200 m from the Sikkim Himalaya.

Genus 2. STYLOPHORUM

Hairy, dwarf-stemmed perennial herb, with stout, woody rootstock. Leaves 7–10 cm long, glaucous, runcinate-pinnatifid, ovate-segmented; petiole alate. Scape slender, 3–4 cm long. Flowers yellow; sepals 2, hairy, cornutely produced at tip; petals 4; ovary 1-celled; style erect; stigma 2–4, ascending, papillose inside; fruit a capsule.

Stylophorum lactucoides blooming in early July at elevations of 3000–3600 m.

Family 3. FUMARIACEAE

Genus 1. CORYDALIS

Erect (not exceeding 40 cm in height) or decumbent and also prostrate, much branched and tufted annual herb, with perennial rootstock crowned by dry leaf bases. Leaves highly variable, generally lobed, narrow, linear-lanceolate, sometimes orbicular decompound, bipinnate. Flowers white, blue, pink or also yellow, in racemes; bracts linear-oblong; sepals 2; petals 4, the outer 2 petals dissimilar, the inner 2 petals clawed; stamens 6, diadelphous; ovary 1-celled; stigma 2, small; fruit a capsule.

A number of species are common in the upper reaches of the forest and subalpine zones of the Himalaya; about ten species grow at elevations between 3600 and 5200 m in the Northwest Himalaya. *Corydalis sibirica* may be considered as typical of the region.

Genus 2. HYPECOUM

Much branched, tufted, procumbent, glaucous, annual herb, about 15–20 cm tall. Leaves 5–10 cm long, pinnatifid. Flowers yellow or pinkish; sepals small, 2; petals 4, the outer 2 petals flat and the inner one 2-lobed;

stamens 4; ovary 1-celled; style short, stigma 2; capsule septate.

Hypecoum leptocarpum is found occasionally in bloom during July–August at elevations of 3000–5200 m.

Family 4. CRUCIFERAE

This is one of the dominant families, representing perhaps the development of steppes derivatives. Many plants grow typically in subarid and arid localities, particularly at extreme elevations and on the inner ranges. About two dozen species are known at present under eighteen genera from the Northwest Himalaya.

Genus 1. ALYSSUM

Hoary, much branched, tufted dwarf herb. Leaves linear, entire, 2.5 cm long. Flowers yellow or white, ebracteate, in racemose corymbs; sepals short; petals clawed, obovate, twice as long as sepals; pod about 8 mm long, flattened, rounded or elliptic, with membranous septum; seeds 2, flat.

The commonest high altitude species from the Northwest Himalaya, *Alyssum minimum*, is in bloom during July–August at elevations of 4200–5200 m.

Genus 2. ARABIS

Prostrate or sub-erect and tufted, stellately hairy, dwarf perennial, usually 10–20 cm tall. Leaves obovate, spatulate, entire or dentate, sometimes also pinnatifid; radical leaves petiolate; cauline leaves sessile. Flowers white, pink or pale rose coloured; sepals with sparse stellate hairs, lateral sepal usually saccate; petals clawed, twice as long as sepals; pod 2.0–3.5 cm long, falcate, flat, keeled or smooth, with membranous septum; stigma subsessile; seeds uni- or biseriate.

Arabis tibetica, easily recognized by its long pods, broad but not basally saccate sepals, blooms generally late in April or early in May to June–July, at elevations ranging from 3000 to 5000 m.

Genus 3. BARBAREA
(Pl. XIV)

Tufted dwarf perennial herb, with angled and rugose branches, 20–45 cm tall. Radical leaves petiolate, cauline leaves sessile; obovate, entire or pinnatifid at base. Flowers yellow; petals with broad claw; pod elongate, flat, keeled, 4-angled, 2–3 cm long, with seeds uniseriate.

Barbarea vulgaris commonly in bloom during May–July at elevations of 3000–3300 cm.

Genus 4. Braya

Much branched, tufted, dwarf herb, generally glabrous but often also hoary, particularly when growing at extreme elevations. Leaves mostly radical, elongate-spatulate, dentate. Flowers pinkish to rose coloured, sometimes also paler and white in drier localities; pods ovate, short or also elongate and cylindrical, with seeds usually biseriate.

Braya rosea, with flowers in dense corymbs, at elevations of 4500–5200 m in July; plants grow from 3600 to 5500 m and are in bloom from May to August, depending upon the altitude.

Genus 5. Capsella

Glabrous or hoary, much branched dwarf or spreading and decumbent annual herb, with perennial rootstock. Radical leaves pinnatifid or even entire, depending on age of the plant; cauline leaves spatulate or linear, entire. Flowers white, in axillary racemes; pods obcordate, oblong-ovate, flattened, with numerous biseriate seeds.

Some species of this common genus are abundant even in the upper reaches of the forest, immediately below timberline, as well as in the subalpine zone of the Northwest Himalaya.

Capsella bursa-pastoris occurs from subalpine zone up to elevations of almost 5900 m, sparsely and isolated in small pockets. Blooming extends from April to almost August, depending on altitude.

Genus 6. Cardamine

Generally glabrous or sometimes sparsely pubescent, erect, perennial herb, usually not taller than 30 cm. Leaves ovate-lanceolate, crenately lobed or pinnatifid. Flowers white, pinkish or rarely violet coloured and sometimes even yellow; sepals equal to base; petals clawed; pods elongate, compressed, narrowed both at base and tip, valves with midrib, septum membranous; stigma simple or bifid; seeds flat, uniseriate.

Cardamine macrophylla, rather variable plant, blooms in May–June at elevations of 2800–3600 m.

Genus 7. Cheiranthus

Hoary and at higher elevations generally woolly, tufted, dwarf herb, 5–15 cm tall, with rootstock covered by the withered petioles. Leaves elongate-oblong, sometimes spatulate in older plants, rarely coriaceous; entire or more usually dentate and sometimes also obscurely lobed at higher elevations. Scape about 20 cm tall. Flowers yellow, but sometimes pink; sepals erect, sometimes also pink coloured like petals, lanceolate and wide, one sepal saccate at base; petals clawed, about twice as long as sepals; pods glabrous, long, 4-angled, flat, with membranous septum.

Cheiranthus himalayensis, with its pink flowers, is readily separated from *Parrya*, by its 4-angled pods, broader sepals of which one is saccate basally; the plants generally bloom about June–August at elevations from 3600 to 5200 m.

Genus 8. COCHLEARAIA

Stemless or prostrate dwarf, usually fleshy annual herb, with perennial underground parts. Leaves elongate-lanceolate, oblong-ovate, dentate or pinnatifid or sometimes entire in young specimens; petioles long, often flattened. Flowers white, but sometimes also yellow or pale violet coloured, in short stunted racemes or more usually corymbs; sepals persistent, spreading; petals clawed; pods 5–8 mm long, subglobose or obliquely oblong, somewhat flattened, pubescent, with biseriate seeds (3–5), rarely also uniseriate, flat, not winged.

Cochlearia himalaica is a prostrate dwarf, fleshy plant, with oblong-ovate and long petioled leaves, found in bloom in July–August at elevations of 4000–5000 m.

Genus 9. CHORISPORA
(Pl. XVI)

Tufted dwarf herb, with glandular hairs. Leaves oblong-lanceolate, dentate, often pinnatifid at higher elevations. Flowers yellow or pale pinkish, in ebracteate racemes; pods twisted, cylindrical, indehiscent.

Chorispora sabulosa with pale yellow flowers in June at 4900 m, *C. sibirica*, *C. elegans* and *C. tenella* with pink flowers in May at elevation of 3600 m [99].

Genus 10. CHRISTOLEA

Decumbent or much branched dwarf and tufted, glabrous or pubescent, perennial herb. Leaves obovate, fleshy, dentate apically. Flowers yellow, ebracteate; pods elongate, with flat biseriate seeds.

Christolea crassifolia blooms in late summer at elevations of 3600–5900 m and also occurs sparsely at 6000 m.

Genus 11. CRAMBE

Stout, hairy herb with stout rootstock. Radical leaves cordate, with stout petioles; cauline leaves ovate, dentate, with short petioles. Flowers white, in ebracteate racemes; pods indehiscent, articulate, the upper segments containing single seeds and the lower segments forming empty, seedless stalk.

Crambe cordifolia flowers in June at elevations of 3000–4000 m.

Genus 12. DRABA

Prostrate or spreading, much branched, tufted dwarf perennial herb, generally not exceeding 10–40 cm in height and densely clothed with stellate hairs, though occasionally also glabrous. Leaves linear, densely rosulate, oblong-spatulate, entire, 5–25 cm long. Scape 15 cm long, naked; flowers yellow, white, sometimes pink coloured; ebracteate racemes or in corymbs; sepals hairy, with scarious margin; petals clawed, twice as long as sepals; pods short, compressed or turgid, oblong-elliptic or ovate, sometimes slightly curved or also twisted, with membranous septum; seeds biseriate, compressed, 4–10 in each pod; pods shorter or longer than pedicel.

The general hoariness and the short pod, with biseriate seeds usually suffice to identify this plant. The species common at elevations of 3000–5500 m include *Draba alpina*, *D. altaica*, *D. incana*, *D. lasiophylla*, *D. stenocarpa* and *D. tibetica*

Geuus 13. ERYSIMUM

Hoary herb with a number of branches arising in a dense tuft from a common rootstock and about 20 cm tall. Leaves elongate-oblong, entire or sinuately dentate. Flowers yellow or orange-yellow, ebracteate; sepals erect, equal; petals clawed; pods elongate, narrow, compressed, 4-angled or cylindrical, with membranous septum, short style or long; stigma bilobed and capitate; seeds numerous, uniseriate.

Erysimum altaicum, a short erect plant with yellow flowers, blooming in May–June or early July, at elevations from 1500 to 4500 m.

Genus 14. EUTREMA

Generally glabrous, stemless perennial or with several tufted branches arising from base and with stout rootstock. Leaves spatulate, entire, sometimes dentate above the middle; radical leaves with long petioles; cauline leaves sessile or subsessile. Flowers white, in corymbs; sepals erect; pods elongate, cylindrical, style short, seeds wingless and 1–2-seriate.

Eutrema primulaefolium with white flowers blooming in June at elevations of 3600–3900 m.

Genus 15. Iberidella

Glabrous, much branched herb, 60–100 cm tall. Leaves generally alternate, rarely opposite; the upper leaves subamplexicaule. Flowers white or pale pink; sepals erect, lateral sepal saccate at base; petals equal; pods flat, oblong-lanceolate, acute, valves costate, not winged, septum membranous, style elongate, seeds 2–6.

Apparently sparse in the Northwest Himalaya; a damaged specimen, doubtfully placed here.

Genus 16. Parrya

Generally stemless or also much branched, tufted and densely woolly dwarf annual herb, with stout perennial woody rootstock, covered by old spongy petioles. Leaves linear, spatulate, entire or dentate or pinnatifid. Scape woolly, erect, 15–30 cm long or even shorter. Flowers generally pink, but sometimes also white and occasionally in corymbs; sepals linear, erect, sometimes woolly and spreading; petals spatulate; stigma sometimes sessile; pods thin, long, flattened, with flattened seeds.

Parryaexcapa at 4500–5500 m and *P. lanuginosa* with masses of delicately pink coloured flowers at 6100 m. Swan [112] reports this species to be in bloom in October at 5600–6100 m on Mt. Makalu.

Genus 17. Sisymbrium

Generally glabrous or often hoary, much branched, tufted dwarf herbs, not over 40 cm tall and with woody rootstock. Leaves 2.5 cm long, obovate-lanceolate, sagittate, lower leaves shortly petiolate, upper leaves usually sessile, dentate, lobed or pinnatifid. Flowers yellow, white or sometimes pink, with or without bracts; lateral sepal saccate; petals clawed; pods slender, cylindrical or flattened, trinervate, with membranous septum and uniseriate seeds.

Sisymbrium mollissimum blooms during May–July at elevations of 3000–4500 m.

Genus 18. Thlaspi

Tufted dwarf herb, hardly 30 cm tall or sometimes also decumbent. Radical leaves oblong-ovate, cauline leaves cordate, often dentate. Flowers white; pods short, rounded or cordate, sometimes obcordate and flattened, with membranous septum; seeds 2 or more.

Thlaspi arvense ascends from the plains up to 4200 m and *T. alpestre* is in bloom in May at 3600 m, but stragglers bloom in June at 5200 m.

Family 5. VIOLACEAE

Genus 1. VIOLA
(Pl. VII)

Glabrous or obscurely pubescent, stemless or decumbent or much branched and tufted dwarf herb, with long stout stolons. Leaves reniform, oblong, ovate, dentate, crenate or almost entire, with the base sometimes hastate, 2–5 cm long and with linear stipules. Flowers generally blue or lilac coloured, 1–2 on peduncles, dimorphic; one form with large petals and the other with small or no petals; sepals lanceolate or short, produced at base; petals erect or spreading, the lower petal largest, spurred or saccate; anthers connate; style clavate or truncated; stigma obtuse, lobed or cupulate; fruit 3-valved capsule.

Viola patrinii sometimes ascends to the upper reaches of the alpine zone and is usually found up to about 3600–4200 m in the Northwest Himalaya and blooming during June–July; stray examples of stunted plants found at 5000 m in July.

Family 6. CARYOPHYLLACEAE

This family is represented by about fifteen species, which occur mainly in subarid to arid localities, often at extreme high elevations, typically as cushion plants.

Genus 1. ARENARIA

Densely tufted, usually hairy, much branched annual herb and often with crowded leaves and forming mats. Leaves linear-oblong, ovate. Flowers white or pale pink coloured; sepals 5, petals 5, stamens 5–10, ovary 1-celled, style 3–4; capsule short, with compressed seeds.

Arenaria serpyllifolia is a much branched and softly pubescent plant, with almost sessile, ovate leaves and leafy bracts and *A. kashmirica* (Fig. 17) in bloom during June at 3300–3900 m. According to Swan [112], *Arenaria* grows at 5200 m on Mt. Makalu.

Genus 2. CERASTIUM

Erect, pubescent shrub or tufted procumbent herbs of highly variable size. Leaves glabrous with alternating patches of hairy lines, small, oblong-lanceolate, sessile. Flowers white; bracts with scarious margin; sepals elliptic-oblong, about half the petal, usually 5; petals 5 or absent, bifid; stamens 10 or sometimes only 5; ovary 1-celled; style 3–5; fruit a capsule.

Cerastium cerastoides (Fig. 18) and *Cerastium dahuricum* common at 3300–5100 m.

Genus 3. GYPSOPHILA
(Pl. XII)

Hoary, tufted, decumbent annual herb, with stout, perennial, woody rootstock. Leaves linear or obovate-spatulate, obtuse, ciliate; radical leaves

Fig. 17. *Arenaria Kashmirica.*

petiolate; cauline leaves sessile. Flowers lilac-coloured; sepals 5, ciliate, membranous, with 5 veins; petals 5, entire or notched at tip, spatulate, with 3 veins; ovary 1-celled, style 2–3; fruit 4-valved capsule, with numerous reniform seeds.

Gypsophila ceratoides in bloom during May at elevations of 3000–3900 m.

Genus 4. Lychnis

Hoary tufted, dwarf-stemmed and often much branched, spreading herb, not over 15 cm tall. Leaves linear-lanceolate to spatulate, glandular-pubescent or also sometimes glabrous. Flowers white to pale red; calyx inflated, with usually 10 green veins; petals small; ovary 1-celled; style 5; fruit a turgid capsule.

Lychnis apetala, with pink coloured flowers, collected at elevations between 3000 and 5900 m.

Genus 5. Sagina

Much branched, procumbent or tufted dwarf herb. Leaves connate basally, ciliate, exstipulate. Flowers green, solitary; sepals 4–5; petals 4–5–0; ovary 1-celled; stamens 4, 5, 8 or 10; style 4–5.

Sagina procumbens found at elevations of 2500–4500 m.

Genus 6. Silene

Densely tufted, pubescent or viscid annual, with numerous short branches from perennial woody rootstock and generally not exceeding a height of 40 cm. Leaves linear-lanceolate; radical leaves spatulate-lanceolate, acute, obtuse or also acuminate. Flowers yellow to brown, sometimes white or also pale red; calyx inflated, sepals 5 with 10 or more veins, sometimes scarious; petals 5, entire or bipartite or laciniate; stamens 10, 5 usually adnate to petals; ovary incompletely 3-celled, with gynophore; styles 3 [5]; fruit a capsule, with reniform seeds.

Fig. 18. *Cerastium cerastoides*.

Though a number of species like *Silene conoides*, *S. graminifolia*, *S. repens* and *S. wallichiana* are recorded from the Himalaya, only *Silene moorcraftiana*, with pale red flowers, has been found in bloom during May at elevations of about 3600 m in the Northwest Himalaya.

Genus 7. Stellaria

Procumbent, decumbent, suberect, dwarf, glabrous or glandular-pubescent herbs, not over 60 cm tall, sometimes with tufted branches or also with stout branches. Leaves sessile, oblong, ovate or lanceolate. Petals 5, stamens 10 with glandular disc; ovary 1–3 celled, style 3; fruit a capsule with compressed seeds.

Stellaria media and perhaps also *S. graminea* (?) grows from 3000 to over 6200 m. According to Swan [112], *S. media* occurs at 6200 m on Mt. Makalu.

Genus 8. Thylacospermum

Remarkably hemispherical cushions of greatly tufted dwarf herb. Leaves minute, imbricate. Flowers solitary; calyx obconical; petals 4–5, obovate, small; stamens 8–10; ovary 1-celled; style 2–3; capsule coriaceous.

Thylacospermum rupifragrum occurs at elevations of 4500–5500 m.

Family 7. TAMARISCACEAE

Genus 1. Myricaria

Dwarf bush of fastigiate and stunted shrub. Leaves small, linear, sessile, tufted. Flowers pinkish; stamens 10, alternately long and short, monadelphous; ovary with 3 sessile stigmas; bracts ovate; sepals free or connate.

Common in arid localities, particularly on the inner ranges, at elevations from 3000 to 4500 m.

Family 8. GERANIACEAE

A small family, with species typical of grassy meadows and relatively moist localities.

Genus 1. Biebersteinia

Tufted herb, with glandular pubescent hairs and perennial rootstock. Leaves stipulate, linear, pinnatisect. Flowers yellow, regular, in racemose panicles; sepals 5, imbricate; petals 5, orbicular, imbricate; stamens 10;

ovary 5-celled; style connate into capitate stigma; carpels separate when ripe to give the appearance of fruits of Potentilla.

Biebersteinia emodi, with yellow flowers in bloom during June–July, at elevations of 4000–5500 m.

Genus 2. Erodium

Spreading herb, with the branches distinctly swollen at nodes. Leaves opposite, rarely alternately smaller. Flowers regular, axillary; sepals 5, imbricate; petals 5; stamens 5, with 5 alternating staminodes; ovary 5-celled, 5-lobed, with long beaks; styles 5; capsule 5-lobed, 5-celled with the carpels indehiscent and separating when ripe.

Erodium tibeticum and *E. cicutarium* occur at elevations of 3000–4500 m.

Genus 3. Geranium
(Pl. X)

Prostrate or much branched, spreading, hoary, dwarf perennial herb, with stout rootstock. Leaves opposite or alternate, orbicular, 3–9-partite, the segments often rhomboid or acute; stipules present. Flowers purple-blue, 1–2 in axillary peduncle, bracteate, regular, large; sepals 5; petals 5, imbricate, hypogynous; stamens 10 (sometimes 5 with 5 staminodes); ovary 5-lobed, 5-celled, beaked; style 5; carpels separate when dry.

Geranium collinum and *G. pratense* bloom from May to July at elevations from 2500 to 4500 m.

Genus 4. Oxalis

Pilose, stemless sour herb, with creeping rootstock. Leaves radical, 3-foliate, leaflets obcordate. Flowers axillary, usually pink; sepals 5, imbricate; petals 5, contorted; stamens 10, free or connate basally; ovary 5-lobed, 5-celled; style 5, with terminal capitate, bifid stigma; capsule loculicidal, dehiscent.

Oxalis acetosella may perhaps be considered as one of the high altitude species and occurs from 2500 to 3600 m.

Family 9. LEGUMINOSAE

This important family is represented by about twenty-five species, which are mostly characteristic of arid steppes and occur only sparsely on grassy meadows. The typical high altitude species from the Northwest Himalaya belong to the genera *Astragalus*, *Caragana*, *Guldenstaedtia*, *Melilotus*, *Ononis*, *Parochaetus*, *Oxytropis*, *Thermopsis* and *Trifolium*. A few others, doubtfully placed specifically, like *Hedysarum cephalotes*, *Lotus corniculatus*, *Medicago falcata* and *M. minima* and an unidentified species of *Onobrychis* (Fig. 19) have been sparsely observed at high altitudes, particularly on the inner ranges.

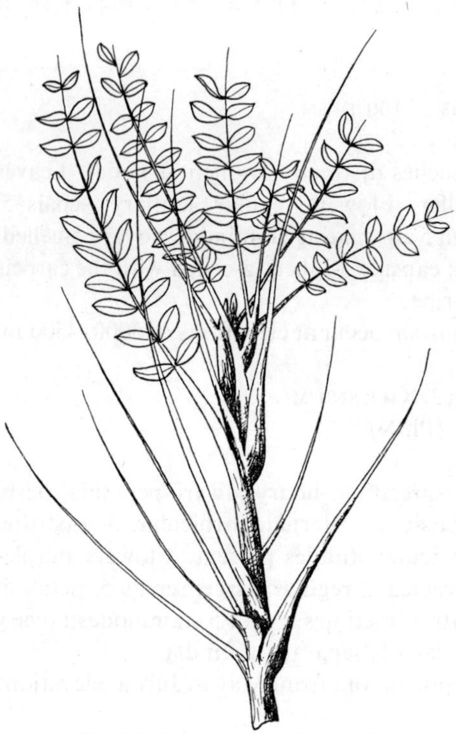

Fig. 19. *Onobrychis echidna* (?).

Genus 1. ASTRAGALUS

Herbs or dwarf shrubs. Leaves pinnate; rachis with leaflets or spiny tipped. Calyx tubular or campanulate; corolla exserted; stamens diadelphous; ovary sessile or stalked; stigma capitate; legume oblong, turgid.

The following species have been identified from the region: *Astragalus alpinus*, *A. densiflorus*, *A. heydei* and *A. tibetanus* (Fig. 20) occurring at elevations from 3000 to 5000 m; *A. heydei* is a dwarf plant that forms densely silky matted growths, about 2–3 cm thick on the ground at 3900–4200 m; *A. tibetanus* is a grey-coloured dwarf plant at 3000–4200 m. All species occur in arid localities.

Genus 2. CARAGANA

Dwarf shrubs. Leaves with rachis and stipules spiny, sometimes the rachis suppressed; leaflets 4-digitate, oblanceolate or oblong-oblanceolate. Flowers sessile, solitary, with pungent, trifid spines, reddish-yellow; calyx campanulate and obliquely attached on pedicel; corolla exserted; stamens diadelphous; ovary sessile; stigma terminal; legume turgid.

Caragana pygmaea is a tufted dwarf, with woody creeping branches, without rachis and digitately 4-foliate leaves and orange flowers, found in subarid and arid localities, from 3000 to 5500 m.

Genus 3. GULDENSTAEDTIA

Spineless perennial, with odd-pinnate and silky leaves; leaflets obovate-obcordate. Flowers solitary or umbellate; calyx campanulate; corolla exserted; stamens diadelphous; ovary sessile; stigma capitate; legume turgid.

Guldenstaedia himalaica is a perennial with woody rootstock, about 2–5 cm tall and leafy stipules, fairly common at elevations of 3300–4500 m.

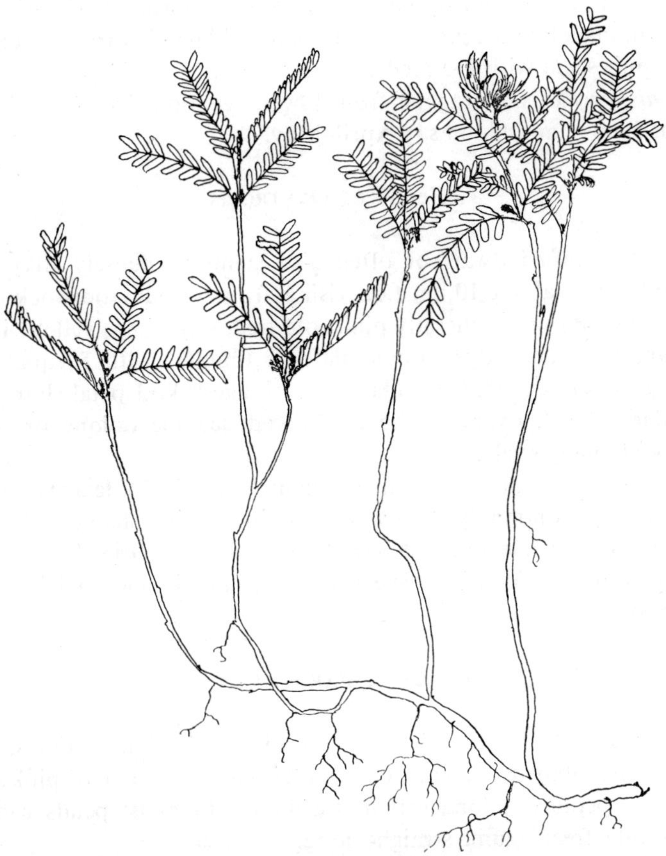

Fig. 20. *Astragalus tibetanus.*

Genus 4. MELILOTUS

Prostrate, much branched annual herb. Leaves pinnately trifoliate. Flowers in long racemes, yellow (also white); calyx campanulate, sepals 5, equal; corolla caducous, free, standard petal narrow; stamens diadelphous, filaments dilated; ovary sessile or stipitate; style filiform with terminal stigma; legume oblong, indehiscent.

Melilotus officinalis and often also *M. alba* occur in grassy meadows at 3000–4000 m.

Genus 5. Ononis

Unarmed, pubescent, dwarf shrub. Leaves pinnately trifoliate; leaflets dentate; stipules large, foliaceous, adnate to the short petiole. Flowers red, in axillary pairs; calyx campanulate, sepals subequal, elongate; petals clawed, standard petal rounded, wing petal oblong-obovate, keel petal curved; stamens monadelphous, filaments alternately dilated; ovary sessile; legume oblong, pubescent with 2–3 seeds.

Ononis hircina is a straggler from lower elevations in subarid localities at about 3000 m and blooms in April–May.

Genus 6. Oxytropis

Generally tufted dwarf or often also stemless, densely silky haired, perennial herbs, about 10 cm tall, rising from woody rootstock. Leaves odd-pinnate. Flowers generally pink coloured, sometimes yellow, in dense or peduncled heads; calyx campanulate or tubular, with 5 equal sepals; corolla generally exserted; standard petal longest; keel petal shortest; stamens diadelphous; ovary sessile or stalked; legume oblong or globose, turgid, with many seeds.

Oxytropis lapponica, a stemless plant with 13–25 leaflets and pink flowers, occurs commonly; *O. mollis* is a silky-white, stemless plant, with up to two dozen leaflets at 3000–3900 m; *O. immersa* (Fig. 21) and a rare yellow coloured unidentified species, was found in bloom in August at 5000–5500 m.

Genus 7. Parochaetus

Prostrate herb. Leaves digitately trifoliate, with long petioles; leaflets almost sessile, obovate-cuneate, entire, glabrous. Flowers pale pink coloured, 1–2 on peduncles longer than the leaves in axils; petals caducous; staminal tube free; legume straight, long.

Parochaetus communis, with long petioles, subsessile leaflets, blooms during May at elevations of 3000–3600 m.

Genus 8. Thermopsis

Perennial herb, with much branched aerial shoot, arising from woody rootstock. Leaves with stipules often foliaceous, generally alternate, trifoliate, sessile; leaflets obovate-cuneate, oblanceolate, glabrous. Flowers yellow, 6–12 in racemes; bracts large, foliaceous; calyx turbinate, sepals lanceolate; petals long clawed; standard petal orbicular, wing petal obovate, keel petal obovate-oblong; stamens free; ovary sessile or stalked, with 6–10 ovules; stigma terminal; legume turgid, oblong, generally straight.

Thermopsis barbata (Pl. XVII & XXI) is a stunted plant found in bloom during May–July at different elevations from 3300–5500 m.

Fig. 21. *Oxytropis immersa.*

Genus 9. TRIFOLIUM

Prostrate herb. Stipules adnate; leaves digitately trifoliate; leaflets obovate, dentate. Flowers white to pink, in axillary globose heads; calyx turbinate, sepals equal 5; corolla and staminate tube adnate; standard petal and wing petals narrow; stamens diadelphous with filaments dilated; stigma oblique; ovary sessile or stalked; legume small, membranous.

Trifolium pratense and *T. repens* on grassy meadows often occurring up to 6000 m.

Family 10. ROSACEAE

This family is represented by about twenty species, belonging mostly to *Potentilla*, which fall apparently under two major ecological groups: the steppes forms and the grassy meadow forms. Most species appear in bloom during May–June at elevations of 3000–3900 m, but sometimes individuals may be found in bloom at 5500 m in July–August. Although the family seems to be but sparsely found at this elevation in the Northwest Himalaya, one species *Sibbaldia tetrandra* is known at an elevation of 6000 m in the Pamirs.

Genus 1. ALCHEMILLA

Densely silky-haired, prostrate annual low-growing plants, with stems covered by dry stipules and perennial creeping rootstock. Leaves orbicular, lobed or deeply divided, alternate, but mostly in dense rosettes, with adnate stipules and sheathing branches. Flowers small, yellow to yellowish-green, in corymbs; bracts numerous; disc adnate to calyx tube; calyx campanulate, lobes 4–5-valvate; stamens 4–5; style basal, with capitate stigma; achenes 1–4, enclosed in membranous persistent calyx.

Alchemilla vulgaris is in bloom during May–June at elevations of 3000–3600 m.

Genus 2. FRAGARIA
(Pl. XXI)

Scapigerous perennial herb, with creeping stolons. Leaves digitately 3–5-foliate, sometimes also pinnate or rarely simple; stipules adnate. Flowers generally white, but sometimes yellow, with 5 bracteoles; calyx persistent, sepals 5, petals 5; stamens numerous, persistent; carpels numerous on convex receptacle; style ventral, persistent; achenes numerous and sunk in large, fleshy receptacle.

Several species are common enough at lower elevations but only one or two appear to bloom during May–July at 3000–3900 m.

Genus 3. GEUM

Erect perennial herb. Leaves radical, crowded together, pinnate; terminal leaflet large; stipules adnate. Flowers white or yellow or red, solitary or in corymbs; bracteoles 5; calyx persistent, sepals 5; petals 5; stamens numerous; carpels numerous; style filiform, elongating after flowering; achenes numerous in dry receptacle.

Geum urbanum is the commonest species at 1800–3300 m; there is another unidentified species with yellow flowers blooming in late June at 4500 m.

Genus 4. POTENTILLA (Pl. XX)

Potentillas are among the commonest plants at high altitudes. Several species are known from the subalpine and on the grassy meadow zones higher. Some species grow exclusively at elevations of 3600–5500 m. The plants have the general appearance of *Fragaria*, but have usually coloured flowers and dry receptacle in fruit. The species are often very difficult to be correctly recognized; the difficulty increases with altitude.

The high altitude forms are typically much branched, spreading, prostrate or also tufted plants, usually more or less densely clothed with silky-white hairs; the branches arise from woody rootstock. Leaves most usually digitately 3–5-foliate, but also pinnate in some species, with the leaflets alternately smaller; leaflets variable, obovate, cuneate-truncate, elliptic-oblong, with stipules adnate. Flowers mostly yellow, red or rarely white, solitary or in cymose corymbs, with persistent calyx, bracteoles 5–4; petals 5 (sometimes flowers unisexual); stamens usually many; carpels many; achenes many on dry receptacle, which has often abundant hairs, concealing the achenes.

Of the several species occurring at high altitudes on the Northwest Himalaya, the following may be considered as common: *Potentilla albifolia* a many branched, pubescent, spreading plant, with solitary flowers in axils, blooming in May–June at 2900–4500 m; *P. anserina* with alternately minute leaflets and yellow flowers at 2800–4800 m; *P. bifurca* with pinnate leaves, but the leaflets not alternatingly smaller, silky haired, flowers yellow at 3000–5100 m; *P. desertorum* with digitately 5-foliate leaves, clothed with glandular hairs, flowers yellow at 2000–3000 m and blooming during April–May; *P. eriocarpa* an almost glabrous species at 3600–4200 m; *P. fruticosa* with dioecious flowers blooming during June at 3900–5000 m; *P. gelida* with digitately 3-foliate leaves, sparsely pubescent, yellow flowers at 3900–5100 m; *P. mooniana* with the leaves interruptedly, alternately smaller, pinnate, the larger leaflets elliptic and the smaller one orbicular and achenes not hidden by hairs at 3000–3600 m; *P. multifida* tomentosely hairy and tufted plant with leaves almost pinnate and yellow flowers at 3000–5200 m; *P. nivea* densely white tomentose dwarf plant, with 3–5 leaflets and yellow flowers at 3000–5100 m; *P. sericea* a much branched and silky haired plant, with yellow flowers at 3000–5100 m; *P. sibbaldi* with digitately 3–5-foliate leaves, densely clothed with white, stiff silky hairs and yellow flowers in bloom during May–June at 3900–4500 m and *P. solossovii* a silky haired plant, with stout woody rootstock and coriaceous leaves, blooming during May at 3000–4200 m.

Genus 5. SPIRAEA

Perennial herbs or shrubs. Leaves alternate, simple or compound; stipules free or adnate or absent. Flowers white or reddish, in axillary or terminal cymes; calyx persistent; petals 4–5; stamens 10, with filaments free or connate below; disc fleshy; carpels 5 or more.

Spiraea canescens, with stout, tomentose branches, small, subsessile, obovate leaves, entire or 3–5-dentate near tip and pale pink coloured flowers, is common from the forest zone upwards to about 3600 m and another species *S. sorbifolia* with 6–9-sessile, lanceolate leaflets and white flowers at 2500–3000 m.

Family 11. SAXIFRAGACEAE

This is a small, but often very conspicuous, member of the high altitude ecosystem.

Genus 1. PARNASSIA

Glabrous, scapigerous, perennial herb. Leaves entire, ovate-oblong or reniform, ovate-cordate. Flowers white or pale yellow; calyx free or adnate to ovary, sepals 5; petals 5; stamens 5, perigynous or almost hypogynous, alternating with 5 staminodes. Ovary 1-celled, style short, stigma 3–4.

Two species *Parnassia laxmanii* and *P. palustris* are generally found at elevations of 2500–4500 m.

Genus 2. SAXIFRAGA
(Pl. VII)

Short-stemmed or densely tufted or decumbent perennial herb, usually about 15–20 cm tall. Radical leaves typically rosulate, entire, reniform or palmately 5–6-lobed; cauline leaves usually simple; bulbils often in axils. Flowers white, yellow or sometimes red, in corymbs; calyx adnate to base of ovary, sepals 5; petals 5; stamens 10,

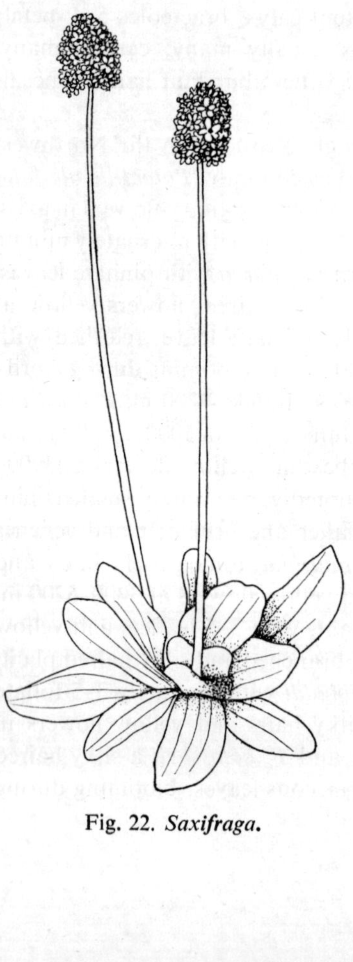

Fig. 22. *Saxifraga*.

sometimes only 5; carpels 2, united at base, stigma lateral, subcapitate; ovary 2-celled (Fig. 22).

About half a dozen species have been observed so far in the Northwest Himalaya, nearly all of which are in bloom late in summer, about end of July. *Saxifraga brunoniana*, with glabrous rosulate leaves and cauline leaves with spinous margin and yellow flowers at 4900 m; *S. cernua* with bulbils in upper axils, solitary flowers, radical leaves reniform, palmately 5–6-lobed and white flowers, at 4200–5100 m; *S. diversifolia* a glandular plant with the radical leaves long-petioled, ovate, subcordate, cauline leaves ovate-oblong and yellow flowers in corymbs with petals about twice the sepals at 3000–5100 m; *S. flagellaris* with cauline leaves linear-lanceolate and pubescent ciliate or even subspinulose and yellow flowers at 4000–5500 m; *S. hirculus* a decumbent plant with almost glabrous leaves, radical leaves petioled, glandular-ciliate and with yellow flowers at 3300–5100 m; *S. ligulata* (Fig. 23) with yellow flowers at 3300–3900 m; *S. sibirica* with 5–7-lobed radical leaves without bulbils and with white flowers in upper axils at 3000–4200 m and *S. stracheyi* with white or pink coloured flowers and undivided obovate glabrous leaves, ciliate or crenate-serrate at 2500–4200 m.

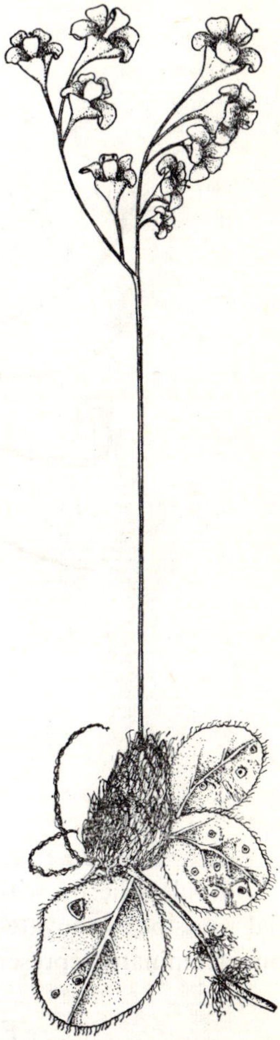
Fig. 23. *Saxifraga ligulata*.

Family 12. CRASSULACEAE

Genus 1. SEDUM
(Pl. XIX, XXIV)

Succulent herbs, with perennial rootstock, ending in a tuft of scales and numerous tufted branches. Leaves mostly rosulate (very rarely opposite), most generally entire though sometimes lacinate. Flowers yellow, pink or red coloured; calyx 5–4-partite; petals 5–4 and free; stamens 10–8, alternately adnate to petals; carpels 5–4, free or slightly united at base; follicles 5–4 with numerous seeds.

Sedum asiaticum with linear leaves and yellow flowers at 3900–4900 m; *S. ewersii* with glabrous, obovate or orbicular opposite leaves and rose coloured flowers at 2800–5100 m; *S. quadrifidum* (Fig. 24) with oblong leaves and red flowers at 3300–5500 m; *S. rhodiola* with yellow flowers, petals longer than sepals in staminate flowers and shorter in pistillate flowers at 3600–5100 m and *S. tibeticum* with red flowers and entire ovate-oblong leaves at 3600–4900 m.

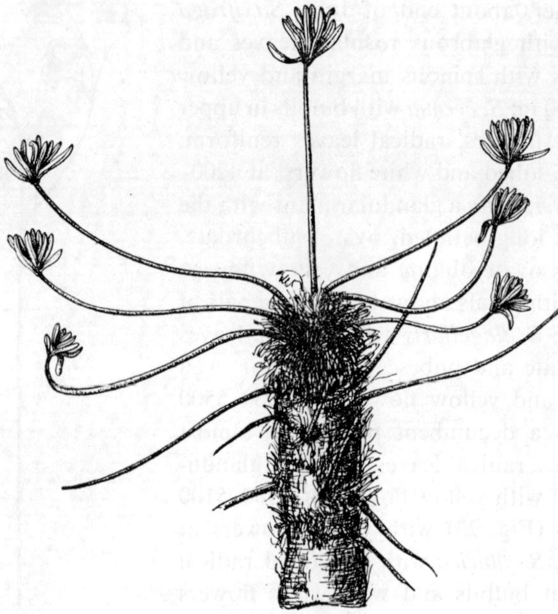

Fig. 24. *Sedum quadrifidum.*

Genus 2. SEMPERVIVUM

Succulent herbs, with the flowering stem of rosette of leaves from perennial rootstock, obovate-lanceolate or nearly linear and cauline leaves oblong acuminate, represented by *Sempervivum acuminatum* at 3000–4500 m.

Family 13. ONAGRACEAE

Genus 1. EPILOBIUM

Often pubescent dwarf herbs, generally 25–30 cm tall. Leaves alternate or more often opposite, obtusely toothed, about 5–7 cm long, obtuse or acute. Flowers pink to purple coloured, solitary and axillary or in spikes; more or less regular; calyx tube slightly produced above the ovary, 4-toothed, deciduous; petals 4; stamens 8, epigynous, 4 short and 4 long; ovary 4-celled, inferior; style with 4 spreading stigma; fruit a 4-celled elongate capsule.

The following five mesophyte species are commonly met with on grassy meadows: *Epilobium angustifolium* with narrow lanceolate leaves at 2500–3900 m, *E. hirsutum*, *E. latifolium* and *E. palustre* at elevations of 2900–4200 m and *E. roseum* at 1500–3600 m.

Family 14. UMBELLIFERAE

The members of this family are conspicuous in semi-arid grassy meadows at elevations of 3900–4500 m, but often stragglers ascend as high as 5500 m. About half a dozen species have so far been found in bloom during May–August in the Northwest Himalaya.

Genus 1. Bupleurum

Low-growing, generally glabrous shrub, with leaves entire; upper cauline leaves subsessile, oblong-ovate, the lower leaves linear-oblong. Umbels compound, with bracts and bracteoles leafy or hairlike or even absent; flowers yellow; fruit compressed.

Bupleurum candolei 2500–3900 m.

Genus 2. Heracleum
(Pl. IX)

Erect herb. Leaves 1–2–3-pinnate. Umbels compound; bracts few, simple or also absent; bracteoles many; flowers polygamous, white or yellow; sepals 0; petals obovate; fruit orbicular-obovate, compressed dorsally.

Two species have so far been observed on grassy meadows: *Heracleum pinnatum* at 3000–4900 m and *H. thomsoni* at 3000–4500 m.

Genus 3. Pimpinella
(Pl. IX)

Stout and tufted-branched, perennial herbs, usually less than 30 cm tall. Leaves 1–2-pinnate, 1–2-ternate, decompound, rarely only toothed. Umbels compound, bracts few or 0, bracteoles linear or absent, flowers pink coloured, hermaphroditic or polygamous or monoecious; sepals 0 or minute, petals emarginate; fruit glabrous, compressed.

Pimpinella saxifraga, with numerous striate branches arising from stout rootstock, is found in bloom during May–June at 3900–4200 m.

Genus 4. Pleurospermum

Stout-stemmed, dwarf perennial, about 20 cm tall. Leaves 1–4-pinnate, the ultimate pinnae ovate, dentate, pinnatifid. Umbels compound, bracts many or few, pinnatifid; bracteoles many, broad, entire, margined white; sepals small or absent; petals obovate, emarginate, white or pink-coloured; fruit elliptic-cylindrical, dorsally compressed.

Among the numerous species found in the Himalaya, *Pleurospermum densiflorum* is readily recognized; it is about 20–30 cm tall, with 3–4 pinnate leaves, the ultimate segments of which are lanceolate-linear; the bracts 5–6 and about 2–5 cm, bracteoles 5–8 and elliptic, entire; the plant is in bloom during May–June at elevations of 3300–4500 m in the Northwest Himalaya.

Genus 5. Selinum

Much branched perennial herb. Leaves pinnately compound. Umbels compound, with numerous rays; bracts many, linear-pinnatifid; bracteoles many, linear, serrate or pinnatifid; calyx without teeth or with linear-lanceolate teeth; petals white to yellow coloured; fruit oval, compressed, with primary ridges alate.

Two species *Selinum papyraceum* at elevations of 2500–3600 m and *S. tenuifolium* at 1800–3900 m are known at present from the Northwest Himalaya.

Genus 6. Trachydium

Often densely hairy, stout stemmed dwarf plants, hardly 5–10 cm tall, with thick grooved branches. Leaves 1–2–3-pinnate. Umbels compound, bracts numerous, foliaceously lobed or pinnate; bracteoles 5–10; calyx entire; petals generally white or sometimes also pink coloured; fruit ovoid, subcompressed, with 5 ridges.

Only one species *Trachydium roylei*, with radical leaves about 7 cm long, 2–3-pinnate with the ultimate segments lanceolate-linear, has been observed at elevations of 3300–5500 m in the Northwest Himalaya.

Family 15. CAPRIFOLIACEAE

Genus 1. Lonicera

Densely branched, dwarf shrub, with scaly buds. Leaves opposite, petiolate or sessile, connate, oblong-elliptic or small and oblong, usually entire. Flowers dioecious, pink coloured, with bracts, bracteoles 2; calyx funnel shaped, with short limbs, five toothed; corolla tube short or long, funnel shaped or campanulate, 5-lobed or 2-lipped; stamens on corolla; ovary 2–3-celled; style slender; stigma capitate; fruit berry.

The following species may be considered as typical of the steppes vegetation of the Northwest Himalaya: *Lonicera hispida* with hispid and ciliate bracts at elevations of 3000–4500 m, *L. microphylla* without bracteoles at elevations of 3300–4200 m and *L. parvifolia* with small oblong leaves at elevations of 3600–4200 m.

Family 16. VALERIANACEAE

Genus 1. VALERIANA (Pl. IX)

Decumbent, stout-branched herb, about 20 cm tall, with stoloniferous rootstock. Radical leaves long petiolate, ovate, entire; cauline leaves pinnatifid. Flowers white or pink coloured, dioecious, in cymose panicles, with small linear-oblong bracts; calyx small, with 5–15 plumose bristles; corolla funnel shaped, petals 5; stamens 3; ovary 3-celled; fruit oblong-lanceolate and compressed, plano-convex and crowned by the pappus-like calyx.

Three species *Valeriana dioica*, *V. dubia* and *V. officinalis* are generally found at elevations between 3300 and 4200 m in bloom during July–August in the North-west Himalaya.

Family 17. DIPSACACEAE

Genus 1. MORINA

Erect herb, about a metre high but generally much shorter. Leaves opposite or most often whorled or in subrosettes, sessile, narrow oblong, spiny-dentate. Flowers pink or yellow, in whorled spikes with spiny bracteoles; calyx 2-lipped; corolla elongate-funnel shaped with the oblique mouth bilabiately 5-lobed; stamens 2 or 4 fertile on the mouth of corolla; achenes free.

Morina coulteriana is sparingly

Fig. 25. *Scabiosa alpestris.*

found at elevations of 3000–4200 m in arid localities.

Genus 2. Scabiosa

Hairy, much branched herb, with stout perennial rootstock. Leaves opposite, simple or more usually pinnatifid, mostly however subradical, sessile or shortly petiolate, oblong. Heads bracteate, pale pink coloured; involucre 2–8-costate; calyx with long hairs; corolla tubular, limb 5-fid; achenes with persistent scarious crowns, 8-ribbed.

Three species are common in dry localities: *Scabiosa oliveri* at 2500–3900 m, *S. alpestris* (Fig. 25) and *S. speciosa* at 3900–4200 m.

Genus 3. Triplostegia

Erect dwarf, glandular-pubescent herb, about 20–50 cm tall, with creeping rootstock. Leaves petiolate, 5 cm long, hairy, dentate or pinnatifid; the uppermost leaves are generally sessile. Flowers pink coloured, in 2–3-chotomous cymes, with narrow bracts, small, sessile, with 2 opposite and narrow bracts and 4 bracteoles; calyx small, minutely toothed; corolla funnel shaped, 5-lobed; stamens 4; stigma capitate; achenes short-beaked and with 4 ribs.

Triplostegia glandulifera occurs at elevations between 2500 and 4200 m.

Family 18. COMPOSITAE

This family is perhaps one of the most characteristic group of arid-frigid steppes components, generally dominant at elevations of 3900–5500 m. Most species are common in rocky localities. There is great abundance both of genera and species and of individuals. We know at present over eighty species, contained in some thirty genera, of which the following may be considered as typical: *Allardia*, *Anaphalis*, *Arctium*, *Artemisia*, *Aster*, *Carduus*, *Centaurea*, *Cnicus*, *Cousinia*, *Cremanthodium*, *Crepis*, *Doronicum*, *Echinops*, *Erigeron*, *Gerbera*, *Helichrysum*, *Hieracium*, *Inula*, *Jurinea*, *Lactuca*, *Leontopodium*, *Saussurea*, *Scorzonera*, *Senecio*, *Tanacetum*, *Taraxacum* and *Werneria*. Some other plants like *Myriactis wallichii*, *Sonchus*, etc., are often found at high altitudes in the region. Many species are characteristic of wind-blown areas. The plants are generally in bloom from April through to end of August or even early September. This is a difficult family to study and there is at present much confusion regarding the limits of genera and species.

Genus 1. Allardia

Generally densely woolly, tufted, aromatic, dwarf herbs. Leaves alternate, about 20 cm long, cuneate, 3–5-dentate or 2–3-pinnatifid. Heads terminal, sessile or peduncled, heterogamous, rayed; involucre bracts woolly, herbaceous, with the margin purple and scarious; rayflorets bisexual, uniseriate, with ligule white or lilac coloured; disc florets fertile, limb 5-fid; receptacle flat, naked; anther base truncate; achenes of bisexual florets 5-angled and glabrous; achenes of pistillate florets empty; pappus rigid, brown to reddish coloured, with undulate bristles.

Two species *Allardia glabra* (?) and *A. vestita* are found in bloom during July at elevations of 4000–5500 m.

Genus 2. Anaphalis

Erect, cottony, woolly dwarf perennial herb, with numerous short and tufted branches, about 7–30 cm long, arising from a woody rootstock. Leaves sessile, linear, elliptic-lanceolate to obovate-spatulate, alternate. Heads small, corymbose, heterogamous, with the pistillate florets outermost or unisexual and subdioecious, disciform; pistillate florets many, filiform, fertile, 2–4-toothed; bisexual florets usually sterile, tubular, 5-fid; ray and disc florets yellow; involucre campanulate, turbinate or subglobose; bracts multiseriate, scarious, inner bracts with petaloid limb, outer bracts shorter, outermost woolly; receptacle naked; anther sagittate. Achenes small, oblong; pappus of pistillate florets uniseriate, scabrid, free, caducous; pappus of bisexual florets stout apically.

Several species occur both in the plains, in the wooded slopes and some of them ascend above the timberline. Of the high altitude species mention may be made of *Anaphalis arenaria* found at elevations of 2500–3900 m, *A. royleana* a much branched dwarf, about 10–15 cm tall with scabrid leaves at elevations of 3000–4500 m and *A. virgata* a cottony, much branched plant, somewhat taller than the foregoing species ascending up to 5000 m. Some of the species of this genus may be identical with those now placed under *Helichrysum* from the Pamirs.

Genus 3. Arctium

Erect herb about 50–100 cm tall. Leaves radical or alternate, undivided. Heads terminal, solitary, sessile or fascicled, homogamous, with all florets bisexual and fertile; involucre subglobose; bracts multiseriate; receptacle flat, densely bristly; anther base sagittate; achenes oblong, subtriquetrous, many ribbed, glabrous; pappus short, copious.

Arctium lappa with densely cottony leaves is in bloom during May–June at elevations of 2500–3900 m.

Genus 4. Artemisia

Glabrous or tomentose, hoary, aromatic herbs or stunted dwarf shrubs, with woody rootstock. Leaves alternate, entire or more usually serrate or 1–3-pinnatisect. Heads solitary, fascicled, racemose or panicled, never in corymbs, heterogamous or homogamous, disciform; involucre ovoid, subglobose or hemispherical; bracts in few series, the outer bracts scarious and short; outer florets pistillate, uniseriate, fertile; disc florets bisexual, fertile or sterile; receptacle flat or raised; anther base obtuse; achenes minute, elliptic-oblong or ovate, obscurely striate, without pappus.

The typical high altitude species of *Artemisia* form characteristic steppes-associations over extensive wind swept areas. The plants bloom from May almost up to early September. Many species occur in the Himalaya, some of which extend almost from the plains upwards. *Artemisia decurrens* is common at elevations of 4200–4900 m, *A. desertorum* at 3000–5500 m, *A. maritima* a tomentose plant with woody rootstock at elevations of 3000–4200 m and *A. salssoloides* with glaucous leaves found sometimes at 5500 m.

Genus 5. Aster
(Pl. X)

Erect or much branched dwarf herb, with stout, woody and sometimes stoloniferous rootstock; branches generally pubescent or more often tomentose and grooved at higher elevations and most usually not taller than 20–30 cm. Leaves alternate, sessile or subsessile; radical leaves petiolate; linear-elliptic, obovate-spatulate to oblanceolate, entire or dentate; cauline leaves subamplexicaule. Heads solitary, corymbose or sometimes panicled, heterogamous and most usually rayed; involucre bracts in few series, linear, oblong to lanceolate, sometimes membranous, usually however leafy. Ray florets 1–2-seriate, bisexual, fertile; ligules 30–60, elongate, white, blue to pink coloured; disc florets yellow, fertile, bisexual, tubular but cleft into 5; receptacle flat or convex; anther base entire or obtuse; styles flat; achenes usually with 1–3 ribs, flat and scanty or copious pappus.

Several species, most of which are not always easily separated from one another, occur in the Himalaya. The validity of some of these species is a matter of some doubt at present, but we may mention here the following four as important at high altitudes in the Northwest Himalaya: *Aster alpinus* and *A. altaicus* are generally found at elevations of 3000–4200 m, *A. heterochaeta* is a woolly, short stemmed plant at 3000–3500 m and *A. molliusculus* a pubescent dwarf stemmed plant but distinctly taller than the foregoing species and found in bloom from April to end of June at elevations of 2000–4500 m.

Genus 6. Brachyaster

Dwarf herb, with glandular, villous or pubescent hairs; usually 20–50 cm tall. Leaves alternate, oblong-lanceolate or ovate, serrate, lobulate or cuneate, decurrent. Involucre bracts 2–3-seriate, linear, the outer one leafy; heads solitary, axillary or also in leafy panicles; usually heterogamous, rayed; rayflorets white, bisexual, in one to many series, fertile, ligule small and sometimes rose coloured; disc florets yellow, bisexual, tubular, fertile; anther base entire, obtuse; pappus biseriate; achenes slender, obovate.

Brachyaster umbrosa is a much branched, villous dwarf plant in bloom during June at elevations of 3000–3600 m.

Genus 7. Carduus
(Pl. XIV)

Erect thistle, with leaves alternate, decurrent, serrate-dentate, pinnatifid, spinescent. Heads solitary on long peduncles or sometimes also subsessile or more or less clustered, homogamous, all florets bisexual, fertile, white or yellow or reddish coloured; involucre ovate, hemispherical or globose; bracts multiseriate, recurved spiny tipped; receptacle flat or convex and densely bristly; anther base sagittate; achenes glabrous, obovoid, obtusely 4-angled, smooth or with 5–10 ribs, truncate or umbonate, pappus copious, multiseriate, simple or barbellate, deciduous.

Carduus thomsoni occurs at elevations of 3300–4000 m. Some of the species placed at present under *Cirsium* may perhaps belong here.

Genus 8. Centaurea

Tomentose or hoary, moderately branched herb. Leaves petioled, pinnatifid or entire. Heads heterogamous, with cottony and multiseriate involucre, bracts with brown and scarious tips; rayflorets large, blue, yellow or white, uniseriate, neuter; disc florets pistillate, fertile; achenes oblong, compressed or 4-angled; pappus bristly or paleaceous, multiseriate.

Centaurea depressa with blue ray florets occurs sparsely in arid localities up to elevations of 3600 m.

Genus 9. Chrysanthemum

Much branched, viscid, woolly, tufted aromatic perennial herb, with woody rootstock. Leaves alternate, mostly dentate, lobed or pinnatifid. Heads in terminal corymbs, heterogamous, rayed; involucre hemispherical, bracts multiseriate, broad, inner bracts scarious tipped; rayflorets pistillate, uniseriate, fertile, with the ligule spreading, white or yellow or rose coloured; disc florets bisexual, fertile; anther base obtuse, entire; achenes subcylindrical or angled, ribbed or winged; pappus absent or short.

We know of two apparently sparsely distributed species *Chrysanthemum richteria*, recognized by its conspicuous woolliness and *C. tibeticum* typically dwarfed, at elevations of 3900–5100 m.

Genus 10. Cnicus
(Pl. XXIV)

The plants generally look like *Carduus*, but are readily distinguished by the foliaceous outer involucre bracts and the feathery pappus hairs. Several species ascend the Northwest Himalaya from the plains. *Cnicus falconeri*, with cottony leaves, is fairly typical of arid localities at elevations of 2500–3300 m.

Genus 11. Cousinia

Dwarf plant with spreading, decumbent branches. Leaves alternate, sessile, hoary, oblong, dentate, often lobed or pinnatisect and spiniscent. Heads solitary or clustered, homogamous, all florets hermaphroditic, tubular, yellow, white or pink coloured; involucre glabrous, ovate or oblong, with multiseriate bracts and spinous recurved tips; anther base sagittate; achenes glabrous, obovate, with 5 or more ribs, compressed; pappus hair uniseriate, simple.

Some of the species placed at present under this genus may perhaps properly belong to *Cirsium* and vice versa. The only species found commonly at 3000–4200 m is *Cousinia thomsoni*. *C. oreodoxa* (Fig. 26) is sparse at 4500 m.

Fig. 26. *Cousinia oreodoxa.*

Genus 12. Cremanthodium

Scapigerous perennial herb. Leaves largely radical, cordate, reniform, dentate; rarely pinnatifid. Heads solitary on leafless or on 1–2-leaved and bracteate scape, radiate, heterogamous, yellow or pink coloured; rayflorets pistillate, uniseriate, fertile, with spreading ligule; disc florets bisexual, fertile; involucre campanulate or hemispherical; bracts uniseriate, membranous; receptacle flat, naked or pitted; anther base subentire; achenes angled, 5–10-striate, glabrous; pappus hair numerous, white or red coloured.

The plants of this genus generally resemble *Senecio*, with about six species on the Himalaya, mostly in the eastern end. One unidentified species, with reniform coriaceous and pubescent leaves was found in bloom during June at 3000–4900 m in the Northwest Himalaya.

Genus 13. Crepis

More or less hairy perennial herbs, but the hairs are not stellate. Leaves radical or alternate; cauline leaves often amplexicaule, entire, dentate or pinnatifid. Heads peduncled, solitary or in fascicled corymbs, yellow or reddish, homogamous, ligulate; involucre cylindrical or campanulate, bracts multiseriate or imbricate; receptacle flat, but sometimes somewhat concave; achenes fusiform, rarely cylindrical; pappus short or long and silvery white.

The species met with at high altitudes include *Crepis multicaulis* ascending up to 5100 m; *C. sibirica* a stout, erect dwarf and hispid plant about 15–20 cm tall, with fusiform rootstock and leaves 15–30 cm long, oblanceolate, acuminate and lyrately pinnatifid and found at elevations of 1300–3300 m; *C. stoliczkai* at 3300–4000 m and *C. tenuifolia* at 3000–4500 m.

Genus 14. Doronicum

Perennial herbs. Radical leaves long petiolate; cauline leaves widely distant and alternate, amplexicaule. Heads large, long peduncled, heterogamous, radiate, yellow; rayflorets pistillate, uniseriate, fertile and with long ligules; disc florets hermaphroditic, fertile, involucre broadly campanulate or hemispherical; bracts biseriate, herbaceous; receptacle hemispherical, naked; anther base entire; achenes of hermaphroditic florets oblong-turbinate, with 10 or more costa; pappus copious; achenes of pistillate florets absent.

Two species *Doronicum falconeri* and *D. roylei* are generally common at elevations of 3000–4200 m.

Genus 15. Echinops

Densely white tomentose plants. Leaves alternate, pinnatifid and spinous. Heads globose, blue or white, sessile or with short stipitae; all florets hermaphroditic and fertile; involucre oblong, bracts multiseriate, rigid, spinescent, the outer bracts short, the inner ones spatulate and the innermost lanceolate; receptacle small; anther base sagittate: achenes elongate, villous; pappus a crown of many short, free or connate bristles.

Two species are generally found at high altitudes: *Echinops cornigerus* with cottony leaves at 1200–3900 m and *E. echinatus* with cottony leaves and involucre villous or with cobweb like coat and white flowers at 3000–3900 m.

Genus 16. ERIGERON
(Pl. IX)

Glabrous or hairy or also often densely tomentose and woolly dwarf plants, with stout rootstock, and having the general appearance of *Aster*, with rather crowded leaves. Leaves petiolate or sessile, 2–5 cm long, oblong, obovate to spatulate, entire or lobulate. Heads solitary or in corymbs; involucre bracts linear, acute, pubescent; ray florets in many series, with narrow, elongate or short, white or pale pink coloured ligules; disc florets yellow; pappus usually single; achenes flat, silky white.

Several species are common at comparatively lower elevations, but *Erigeron alpinus* is perhaps the most common at elevations of 2500–5500 m.

Genus 17. GERBERA
(Pl. XIII)

Scapigerous herb. Leaves radical, scape sometimes scaled. Heads solitary, heterogamous, radiate but sometimes homogamous and discoid, yellow, white or pink coloured; rayflorets 1–2-seriate, pistillate, fertile or also sometimes absent; disc florets bisexual, fertile, 2-lipped, the outer lip of rayfloret with 3-dentate ligule, the inner 2-partite; involucre turbiniform or campanulate, bracts 2 or multiseriate; anther base sagittate; style arms of hermaphroditic florets obtuse; achenes flat, 5-costate, tip beaked; pappus abundant.

Gerbera kunzeana with oblong or ovate leaves often ascends up to elevations of 3900 m, but is more common within the upper reaches of the forest.

Genus 18. GNAPAHLIUM

Plants of this genus are common enough in the plains of India and some species ascend high in the Himalaya. The typical high altitude species are many branched, short and stout stemmed plants rising from stout rootstock. Leaves alternate or apparently so, but most usually rosulate, linear, woolly and entire. Heads small, usually in corymbs, heterogamous, disciform, not rayed, fertile, outer florets pistillate, 2-many seriate, with 3–4-toothed limb; disc florets hermaphroditic, with 5-toothed limb; involucre ovoid or campanulate, bracts many seriate with scarious margins; anther base sagittate; style arms truncate-capitate in hermaphroditic florets; achenes oblong, obovoid, smooth; pappus uniseriate.

Two species deserve mention here: *Gnaphalium stewartii* a cottony-woolly plant at 3000–3300 m and often also found at 3900 m and *G. thomsoni* also a cottony, stout dwarf common within the upper reaches of the forest and often ascending to elevations of 3000 m.

Genus 19. HELICHRYSUM

Woolly herbs. Leaves alternate, entire. Heads heterogamous, or homogamous, not rayed, often clustered in corymbs; in homogamous heads all florets are bisexual; in heterogamous heads the outer row of florets are pistillate, the inner rows bisexual and fertile; involucre bracts many seriate, scarious, yellow coloured or paler; receptacle flat, often pitted; corolla of pistillate florets minutely toothed and of the bisexual florets 4–5-toothed; anther base sagittate, tailed; style arms of pistillate florets slender and of the bisexual florets capitate; achenes small, cylindrical, angled or somewhat compressed; pappus 1-many-seriate, feathery or dilated above.

Unidentified species found at 3900–4200 m.

Genus 20. HIERACIUM

Scapigerous, dwarf perennial herb. Leaves 7–9 cm long, entire, dentate or sometimes pinnatifid. Heads with glandular hairs, solitary, homogamous, yellow or red, all ligulate; involucre tubular, campanulate, bracts elongate, herbaceous; achenes oblong, cylindrical, with 10–15 costa; pappus hairs 1–2-seriate.

Hieracium umbellatum is found commonly in bloom from April to June at elevations of 1600–3600 m.

Genus 21. INULA

Stemless plants or with numerous dwarfed stems clumped together and springing from woody rootstock and hardly 10 cm tall. Leaves radical, usually rosulate. Heads solitary, corymbose or rarely in panicles, heterogamous, radiate, rarely disciform; rayflorets pistillate, one to many seriate, fertile, ligule long, short or even minute, yellow or white coloured; disc florets bisexual, fertile, yellow; involucre narrow or broad, multiseriate, with the outermost bracts foliaceous and the inner bracts rigid; receptacle flat or swollen, pitted; anther base sagittate; achenes subterete, ribbed; pappus short, 1–2-seriate, smooth, scabrid or bearded.

Inula obtusiloba, with numerous clumped branches rising from woody rootstock and sessile oblong leaves, occurs at elevations of 3000–4200 m and *I. rhizocephala* (Fig. 27) is a stemless dwarf hardly 10 cm tall, with all leaves radical and rosulate, obovate-spatulate, ciliate and lying flat and spreading on the ground and with crowded heads in bloom during May–July at elevations of 3000–4200 m.

Fig. 27. *Inula rhizocephala.*

Genus 22. JURINEA

Stemless or much branched and tufted perennial herb. Leaves entire, dentate or pinnatifid. Heads homogamous, all florets bisexual, pink coloured and similar; involucre multiseriate, imbricate, with the innermost bracts narrow and the outer short; receptacle flat, densely bristly; anther base sagittate; achenes glabrous, 4–5-angled, smooth or with 1–2 ribs; pappus hairs multiseriate, simple, barbellate or feathery.

Two species are known at present. *Jurinea ceratocarpa* is a stemless or much branched plant, with leaves linear-oblong or lanceolate, entire or pinnatifid, occurring at elevations of 3000–4500 m. *J. macrocephala* with oblong-lanceolate leaves is found at 3300–4200 m.

Genus 23. LACTUCA

Glabrous or hispid herb with milky latex. Leaves radical and alternate, entire or dentate, pinnatifid or pinnate; cauline leaves amplexicaule. Heads sessile or pedunculate, in panicles or racemose corymbs, yellow, (rarely purple or blue), homogamous, all ligulate; involucre narrow, bracts few seriate; receptacle flat, naked; achenes compressed ovoid-oblong, beaked, 3-ribbed, with the middle rib stronger than others; pappus copious, simple, white.

The following three species have been commonly found at high altitudes in the Northwest Himalaya: *Lactuca orientalis* a spinescent and dichotomously branched plant with small cauline leaves and yellow ligulate florets found in bloom during May–July at elevations of 3000–4200 m; *L. scariola* is a glabrous plant at elevations of 1200–3600 m and *L. tartarica* also a glabrous plant found in arid localities at elevations of 3600–4900 m.

Genus 24. LEONTOPODIUM

Tufted, woolly perennial herb. Leaves entire. Heads yellow, small, sessile in involucrate clusters, disciform, unisexual or bisexual, with the pistillate florets outermost; fertile; bisexual florets tubular and sterile; involucre campanulate, with many seriate bracts, the inner bracts scarious and acute, the outer bracts smaller and woolly; receptacle naked, convex; anther base sagittate; achenes small, oblong, subterete, not ribbed; pappus slender, uniseriate.

Leontopodium alpinum blooms during June–July at elevations of 3000–5500 m. Swan [111] records the genus at elevations of 5100–5200 m in the Central Himalaya.

Genus 25. SAUSSUREA
(Pl. XIV)

Stemless plants or simple or much branched and densely tufted dwarf or rarely spreading perennial herbs, usually tomentose or even cottony haired. Leaves unarmed, alternate, entire or dentate or pinnatifid or pinnatisect. Head crowded together on the dilated tip of stem, shortly peduncled or sessile, often in corymbs, pink or blue coloured, homogamous, with all florets bisexual and similar; involucre ovate, oblong, globose or hemispherical; receptacle flat or convex, densely bristly; anther base sagittate; achenes glabrous, oblong, 4-ribbed, smooth or rugose, truncate or crowned with thick disc; pappus hairs 1–2-seriate, the inner series feathery, stout basally.

There are several species at high altitude in the Northwest Himalaya;

according to the observations of Swan [111, 112], the genus is generally more common at elevations of 4800–4900 m in the area of Mt. Makalu, but the plant is found even at 5100–5500 m. The common species of the Northwest Himalaya include *Saussurea atkinsoni* at 3000–3900 m; *S. bracteata* a dwarf and scaberulous plant at 4200–5100 m; *S. gossypiphora* with dense yellow or white woolly hairs at 4200–5100 m; *S. kunthiana* a stemless dwarf plant with spreading pinnatifid leaves covered by white tomentum beneath at elevations of 4200–4900 m; *S. schultzii* at 3000–4500 m; *S. sorocephala* (Fig. 28) a dwarf and densely tufted stemless (or with very short stem), woolly or cottony plant at 4200–5500 m; *S. stoliczkai* a dwarf and densely tufted cottony-woolly plant at elevations of 3900–4900 m; *S. subulata* also a densely tufted dwarf plant at 4500–5500 m and *S. taraxifolia* a cottony haired spreading dwarf with rootstock at 3600–5100 m.

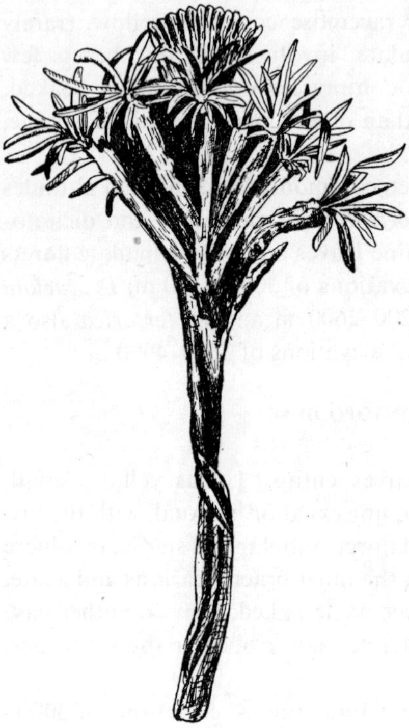

Fig. 28. *Saussurea sorocephala.*

Genus 26. SCORZONERA

Generally scapigerous herbs. Leaves alternate or all radical, entire or lobed or pinnatifid or pinnatisect, elongate and narrow in some forms. Heads large, with long peduncles, yellow, homogamous, all ligulate; involucre cylindrical, campanulate, with multiseriate bracts; receptacle naked or pitted; achenes long, subcylindrical, many ribbed; pappus multiseriate, feathery.

Of the several species found in the Himalaya, *Scorzonera divaricata* is a much branched plant found in bloom during June–July at elevations of 3000–4500 m.

Genus 27. SENECIO

Dwarf shrubs or much branched and spreading herbs. Leaves radical or alternate, entire or divided. Heads solitary or in corymbs or racemes; heterogamous, yellow; rayflorets pistillate, fertile, ligulate; disc florets hermaphroditic, fertile; involucre bracts 1–2-seriate; receptacle flat or convex, naked or pitted; anther base obtuse or auricled; achenes sometimes dorsally compressed, 5–10-ribbed; pappus copious, soft, white or sometimes barbellate.

Several species occur at high altitudes and extend from about 2800 to 5000 m. *Senecio amplexicaulis, S. coronopifolius, S. graciliflorum, S. jacquemontianus, S. retusa, S. thomsoni* and *S. tibeticus* may be generally found in bloom during summer.

Genus 28. Tanacetum

Hoary, tomentose or woolly and densely tufted, slender branched perennial herbs, with woody rootstock. Leaves pinnatisect. Heads corymbose, heterogamous or also homogamous, disciform, yellow; involucre short, multiseriate, bracts with scarious margin, the outer bracts smaller; outer florets absent or when present pistillate, uniseriate, fertile, cylindrical or compressed, 2–3-fid; disc florets bisexual, fertile, stout; receptacle flat or convex; anther base obtuse; achenes compressed, 5-angled, ribbed or triquetrous in rayflorets; pappus annular.

The common high altitude species from the Northwest Himalaya include *Tanacetum artemisioides, T. fruticosum, T. gossypinum* with silky-cottony hairs, *T. senecionis* with woolly stout stemmed dwarf growth and woody rootstock and radical leaves palmately divided, *T. tenuifolium*, and *T. tibeticum* at elevations of 3000–4800 m.

Genus 29. Taraxacum
(Pl. XVI)

Scapigerous plants with milky latex. Leaves radical, sessile, oblanceolate, entire, sinuate, runciate-pinnatifid. Heads solitary on leafless scape, yellow, homogamous, all ligulate; involucre campanulate, bracts herbaceous, the inner uniseriate, outer multiseriate; receptacle flat, naked; achenes oblong, 4–5-angled, 10-ribbed, the ribs muricate above, beaked; pappus simple and unequal.

Taraxacum officinale is in bloom in semi-arid rocky localities during June–July at elevations of 3000–5500 m.

Genus 30. Werneria

The plants of this genus look very much like a short and stout *Senecio*, with creeping rootstock or tuberous roots. Leaves small, elliptic-ovate, entire or coriaceous; radical leaves about 2.5 cm long; cauline leaves about 6 cm long with numerous veins; scape with few short, subsessile leaves and hairy; heads solitary, with yellow ligules; involucre woolly, connate below to form a cup; achenes smooth; pappus abundant.

Werneria nana is the commonest species at 4800 m.

Family 19. ERICACEAE

This family is conspicuous near the timberline. Though well known as the heath zone on tropical-equatorial (especially African) mountains, many species of the family occur on mountains of the temperate zone also. The plants are among the typical shrub elements and belong to the early blooming members of the high altitude ecosystem. We know three genera *Cassiope*, *Gaultheria* and *Rhododendron* from the Northwest Himalaya.

Genus 1. CASSIOPE

Densely tufted, fastigiate, dwarf shrub, about 10–25 cm tall. Leaves small, rigid, densely imbricate, ovate-oblong and often also ciliate. Flowers white or pink coloured, axillary, bracteate; sepals 5, free, glabrous, thick basally, margin scariate, imbricate in bud; corolla campanulate, 5-lobed, short, recurved; stamens 10, basally attached to corolla; filament linear; anther with 1–2 horns; ovary 5-celled, with numerous ovules; capsule globose, depressed at tip, loculicidal, 5-valved; seeds ellipsoid, curved.

Cassiope fastigiata is in bloom from April to May at elevations of 3000–4200 m.

Genus 2. GAULTHERIA

Erect or procumbent or prostrate and much branched hirsute shrub. Leaves subsessile, alternate, persistent, oblong, entire or serrulate-ciliate. Flowers small, red or white coloured, sometimes dimorphic, bracteate, bracteoles 2; calyx ovoid, 5-fid, enlarged and coloured in fruit and succulent; corolla ovoid-tubular, greenish-white or red coloured; stamens 10, with the filaments dilated and pilose; anthers produced upward dorsally 1–2-horned; in dimorphic flowers the stamens small; ovary 5-celled, loculicidal; ovules many; capsule 5-valved from apex; seeds many, minute.

Gaultheria trichophylla is a much branched prostrate plant in bloom during July at elevations of 3000–3900 m.

Genus 3. RHODODENDRON
(Pl. III, XIV)

Several species of this genus are trees and are common in the Himalayan forests; *Rhododendron* is perhaps more common in the Eastern than the Northwest Himalaya. Near the timberline and in areas immediately above the plants are shrubs. The following species appear in subalpine and lower alpine areas and are generally associated with *Betula utilis* and *Juniperus communis*. *Rhododendron anthopogon* a dwarf shrub, about 30 cm tall and densely scaly with elliptic leaves and yellow flowers at elevations of 3300–4900 m; *R. campanulatum* a shrub about two or three metres tall, with elliptic-subobtuse, tomentose-woolly leaves and purple flowers at elevations of 3000–4900 m and *R. lepidotum* a scaly glandular shrub, with obovate to lanceolate leaves and yellow flowers at 2400–4500 m. Most plants are in bloom early in summer, especially during April–May and rarely in June.

Family 20. PLUMBAGINACEAE

Genus 1. Acantholimon

Densely tufted, prickly dwarf herb. Leaves crowded, linear, needle-shaped, spiny tipped, about 2.5 cm long. Scape 2–5 cm long, with many rose-red coloured flowers; bracts usually 3; calyx cylindrical, pilose outside and 10-ribbed, limb white, funnel shaped, scarious; petal oblong, united basally; stigma capitate; fruit an utricle contained inside the calyx.

Acantholimon lycopodioides is typical of arid wind blown localities at elevations of 3300–4500 m.

Family 21. DIAPENSIACEAE

Genus 1. Diapensia

Prostrate or densely tufted dwarf shrubs. Leaves alternate, but usually crowded together, ovate-oblong, entire or dentate, coriaceous, hardly 5 mm long. Flowers at tips of branches, solitary, white or pale reddish coloured; bracts 2–3; calyx persistent, sepals 5, usually broadly ovate; corolla hypogynous, campanulate, persistent, petals 5, spreading; stamens 5, inserted on the throat of corolla; anther subsessile; fruit coriaceous, 3-celled, loculicidal capsule, with numerous seeds.

The commonest species is *Diapensia himalaica* in arid and wind blown localities at elevations of 3000–4500 m.

Family 22. PRIMULACEAE

This is one of the typical high altitude families, with about sixteen species, largely characteristic of glacier-grassy meadows and mesophile to hygrophile localities. The plants are commonly in bloom up to the highest limits of altitude range. Three genera *Androsace*, *Lysimachia* and *Primula* are known.

Genus 1. ANDROSACE
(Pl. XI, XIII, XVIII)

Perennial dwarf herbs. Leaves rosulate or imbricate. Flowers small, white or red; calyx 5-lobed; corolla hypocrateriform, with short tube; stamens 5; ovary ovoid.

There are a large number of species in the Himalaya, most of which are in bloom during June–July. *Androsace lanuginosa* at 2500–3300 m, *A. microphylla* at 3600–3900 m, *A. muscoidea* at 3600 m, *A. sempervivoides* at 3300–4200 m, *A. septentrionalis* at 3600–4900 m and *A. villosa* at 3600–5000 m. Swan [112] has recorded the genus at elevations of 5100–5200 m on Mt. Makalu.

Genus 2. LYSIMACHIA

Erect, procumbent or prostrate creeping herb. Leaves entire, alternate or whorled. Flowers pale pink or rarely yellow coloured, solitary, racemed or even spicate; calyx 5–6-partite; corolla rotate or funnel shaped, 5–6-partite; stamens inserted on base of petal; ovary subglobose; capsule 5-valved.

Species of this genus are generally not found much above 3000 m in the Northwest Himalaya.

Genus 3. PRIMULA
(Pl. I, VIII, XXII)

Scapigerous perennial herbs, with horizontal woody rootstock. Leaves with long or short petioles, sometimes densely sessile, obovate-spatulate, elliptic-ovate-spatulate or oblong or broadly or narrowly oblanceolate, entire, obscurely crenate, dentate. Scape stout and sometimes also slender, 10–35 mm long; flowers usually pink, rose, red to deep purple, pale blue or yellow, rarely also white coloured, in whorls or in umbels, capitate and sometimes dimorphic; one type of flower with short stamens and long stigma inside the corolla tube and the other type with long stamens and short stigma; bracts short; calyx cleft into 5; corolla hypocrateriform or also funnel shaped, petals 5; stamens 5; ovary globose; fruit 5-valved capsule.

Of the several species occurring in the Himalaya, the following may be grouped according to the colour of their flowers as below: 1) Species with yellow flowers: *Primula floribunda* (Pl. XXII), *P. reticulata* and *P. stewartii*; 2) species with rose or lilac coloured flowers: *P. rosea* and *P. denticulata*; 3) species with deep purple flowers: *P. farinosa* and *P. sibirica*; 4) species with blue flowers: *P. tenella* and 5) species with white flowers: *P. involucrata.*

Primula denticulata with stout rootstock, sparingly mealy and leaves do not develop fully until flowering, fleshy scales, leaves obovate-spatulate, obtuse or often minutely denticulate, scape about 25 mm long, stout, and flowers in depressed capitate head in bloom during June at elevations of 2500–3900 m. *P. farinosa* occurs at elevations of 3600–5100 m. *P. floribunda* has glandular-pubescent and elliptic leaves with winged petiole and coarsely dentate and occurs commonly at elevations of 1000–2000 m, but sometimes also at 3000 m. *P. involucrata* resembles *P. sibirica*, but is readily separated by its longer leaves, which are also coriaceous, generally entire but also obscurely crenate and scape long and occurs at elevations of 3600–4500 m. *P. reticulata* is glabrous with long petioles, leaves oblong-cordate, crenate, glaucous and reticulate beneath, scape long, bract linear-oblong and funnel-shaped corolla at 3300–4500 m. *P. rosea* is also glabrous and densely sessile, obovate-oblong, spatulate, oblanceolate, crenulate or dentate leaves, scape with 5 to many flowers at 3000–3900 m. *P. sibirica* is with coriaceous leaves about 2 cm long and long slender petiole, generally elliptic-ovate-spatulate, entire or crenulate and scape with only 3–6 flowers at elevations of 3900–4500 m. *P. stewartii* has glabrous and coriaceous leaves, broadly or narrowly oblanceolate, ovate-spatulate, entire or crenulate or finely denticulate and occurs at elevations of 3600–5000 m. *P. tenella* generally occurs at elevations of about 3600–3900 m.

Family 23. GENTIANACEAE

A small family of about twenty species, largely mesophytes, which bloom during May–July on grassy meadows. Five genera *Exacum*, *Gentiana*, *Jaeshkea*, *Pleurogyne* and *Swertia* may be considered as typical high altitude forms.

Genus 1. Exacum

Small shrub. Leaves opposite, ovate, lanceolate, 3–5-nerved. Flowers blue, purple or white coloured, in terminal or axillary cymes; calyx deeply 4–5-lobed, sepals winged, keeled; corolla rotate, tube globose, petals 4–5, ovate-oblong, twisted to the right; stamens 5–4 on throat of corolla, with short filaments and oblong-ovate anthers; ovary 2-celled; style long; stigma subcapitate; fruit capsule, globose, septicidally bivalved, with numerous seeds.

Strictly speaking, the genus *Exacum* cannot be described as a high altitude plant. *Exacum tetragonum* ascends from the plains (winter weed) up to 2500 m on the Himalaya and may occasionally be found at 3000 m as stunted growths.

Genus 2. GENTIANA

Much branched dwarf herbs, usually with stout rootstock. Leaves opposite, often basally connate. Flowers axillary or terminal; calyx tubular, rarely keeled, sepals 5–4; corolla regular, tubular, campanulate or infundibular, without pits near base, petals 5–4 with folds in between; stamens 5–4 and inserted on middle or below on corolla tube, with the filaments slightly flattened below, anthers oblong, ovate; ovary 1-celled; style short or absent; capsule sessile or stalked, elliptic or oblong, compressed, with numerous seeds.

The following species occur in the Northwest Himalaya: *Gentiana aquatica* at 3900–4800 m, *G. argentea* 2800–3600 m, *G. aurea* with yellow flowers and oblong-elliptic leaves at 3000–4200 m; *G. aurea* and *G. barbata* generally at elevations of 3900 m; *G. borealis* with pink coloured flowers at 3600–4200 m, *G. dahurica*, *G. decumbens* a dwarf plant with stout rootstock at 3300–4500 m, *G. humilis* at 3900–4800 m, *G. pygmaea* at 3000–4900 m, *G. quadrifaria* at 3000 m, *G. riparia*, *G. squarrosa* at 3000–4800 m, and *G. tenella* at 3000–4200 m. *G. moorcraftiana* with blue flowers is perhaps the commonest species that occurs from elevations of about 2800 m to nearly 3900 m and *G. thomsoni* is a much branched dwarf plant at 5100 m. Swan [111] has reported the occurrence of *G. urnula* at an elevation of 6000 m on Mt. Makalu.

Genus 3. JAESHKEA

Erect or decumbent herbs, differing from *Gentiana* in the stamens being inserted in between the petals. Leaves opposite, linear-lanceolate or ovate, with the base connate. Flowers in panicles, blue; sepals lanceolate or orbicular; corolla regular, tubular, petals 5, short, subtriangular; ovary 1-celled; style short; stigma 2-lobed; capsule elliptic, sessile or with a short stalk.

Two species *Jaeshkea argentianoides* at 2500–3900 m and *J. latisepala* at 3000 m are known at present from the Northwest Himalaya.

Genus 4. PLEUROGYNE

Tufted dwarf plants 2–7 cm tall, but sometimes also decumbent in the upper limits of the altitude range. Flowers blue, with green nerves; corolla rotate; stamens at base of corolla; sepals elliptic-linear; ovary 1-celled.

Pleurogyne carinthiaca is commonly found at elevations of 3000–3900 m and some other unidentified species occur at 5500 m.

Genus 5. SWERTIA

Perennial herbs. Leaves opposite. Flowers blue or also white, sometimes with yellow gland; sepals 4–5 and lanceolate; corolla rotate, petals 4–5, twisted to the right; stamens 4, free, inserted near base of corolla; ovary 1-celled; style absent or rarely short, with 2 stigma; capsule sessile, ovate or oblong.

Swertia cordata with sessile leaves 5–3-nerved and cordate base occurs commonly from elevations of 1500 to 3600 m. *S. petiolata* has several tufted branches arising from perennial rootstock, with the lower leaves having long petioles; flowering branch annual; the whole plant is about 40–50 cm tall; flowers blue or white with green tips and veins; generally in bloom in May–June at elevations of 3000–3900 m. *S. purpurescens* is 20–100 cm tall, with oblong or lanceolate leaves 3–1-nerved and 3 cm long, purple flowers which may sometimes be also dark reddish and is in bloom during June at elevations of 3000–3900 m. Some of these species are generally found even within the upper limits of the forest. Most of them are characteristic of grassy meadows, though rarely isolated examples grow in rocky areas. *S. marginata* (Fig. 29) occurs in arid localities at 3900 m.

Fig. 29. *Swertia marginata.*

Family 24. BORAGINACEAE

About twenty species, largely with subarid steppes characters, are known at present from the Northwest Himalaya; most of them are in bloom during May–June.

Genus 1. ARNEBIA

Hispid, diffuse dwarf herb, 7–25 cm tall, with perennial rootstock, staining red. Leaves alternate. Flowers sessile, yellow or white coloured, bracteate; sessile linear; corolla tube slender, longer than the sepals, throat naked, petals 5 and obtuse and spreading; stamens 5 and inserted below the mouth of the corolla; ovary deeply 4-lobed; nutlets 4, ovoid-oblong.

Arnebia tibetana and *A. thomsoni* may be found at elevations up to 3900 m.

Genus 2. Cynoglossum

Hairy, erect, perennial herbs. Leaves alternate, radical leaves petiolate. Flowers ebracteate, blue or pink coloured; calyx deeply 5-lobed, spreading in fruit; corolla tube short, with 5 scales in throat, petals 5; stamens 5 included, anthers small; ovary 4-lobed, with 4 nutlets.

Cynoglossum wallichii is perhaps the typical high altitude species that blooms during June at 3000–3600 m.

Genus 3. Echinospermum

Hispid perennial dwarf herb, about 10–40 cm tall. Leaves alternate, oblong-lanceolate. Flowers small, sessile or with short pedicel, bracteate; sepals 5; corolla tube short, with 5 scales near the mouth, petals 5 and spreading; stamens 5 included; ovary 4-celled; style between lobes; nutlets 4.

Echinospermum barbatum occurs at elevations of 2800–3600 m.

Genus 4. Eritrichium

Strigose, silky-villous, perennial dwarf herb, about 7–12 cm tall. Leaves alternate. Flowers small, blue, but sometimes white coloured; sepals 5; corolla tube short, with 5 scales in throat, petals 5 and spreading; stamens 5 and included; ovary 4-lobed; nutlets 4.

Eritrichium strictum is the commonest species, but another species *E. villosum* occurs scattered in arid localities at elevations from 2500 m to 4900 m.

Genus 5. Lindelofia

Densely hairy, erect dwarf perennial herbs, about 30–60 cm tall. Radical leaves petiolate, 5–10 cm long; cauline leaves alternate, ovate to linear-lanceolate. Flowers blue, pink or almost black, without bracts; sepals enlarged in fruit; corolla tube cylindrical, longer than calyx, with 3 scales in throat, petals obtuse and spreading; stamens 5, anthers large and linear-oblong; ovary 4-celled; stigma small; nutlets 4, depressed.

Lindelofia spectabilis occurs at elevations of 3000–4500 m.

Genus 6. Macrotomia

Hispid, erect dwarf perennial herbs, usually 20–35 cm tall. Leaves alternate, oblong or linear. Flowers pinkish-blue coloured in spikes; sepals 5 and enlarging in fruit; corolla long, naked throat, petals 5 and spreading; stamens 5; ovary deeply 5-lobed; nutlets 4 and ovoid, rugose, trigonous.

Macrotomia benthami, *M. euchroma* (Fig. 30) and *M. perennis* are generally found at elevations of 3000–4200 m in the Northwest Himalaya.

Fig. 30. *Macrotomia euchroma*.

Genus 7. MICROULA

Stemless, pubescent herb. Radical leaves elliptic-spatulate, rosulate, tufted, longer than the inflorescence. Flowers small, sessile, bracteate in dense cymes; calyx campanulate, sepals 5, ciliate; corolla short, with 5 scales in throat, petals 5 and spreading; stamens 5; ovary deeply 4-lobed; stigma capitate; nutlets 4.

Microula benthami occurs at 4500–5100 m.

Genus 8. MYOSOTIS

Hairy perennial dwarf herb, 8–30 cm tall. Leaves alternate, but more usually densely crowded together, 3 cm long. Flowers sessile, ebracteate; calyx deeply 5-fid; corolla blue coloured but changing to red or white later, short, with 5 spreading petals; stamens 5 and included; ovary deeply 4-lobed; nutlets 4 and ovoid-oblong.

Three species *Myosotis caespitosa*, *M. palustris* and *M. sylvatica* have so far been found at elevations of 3000–5100 m.

Genus 9. Onosma

Hispid, silky-white, dwarf herbs. Leaves alternate. Flowers yellow, but sometimes also pink coloured, bracteate, in one-sided corymbs; calyx 5-parttite, sepals acute; corolla tubular or ventricose, throat dilated or constricted, petals 5 short; stamens 5, inserted near the middle of corolla tube, anthers lanceolate, included or semi-exserted; ovary deeply 4-lobed; nutlets 4, ovoid, smooth or tuberlate.

Two species *Onosma bracteatum* and *O. gmelini* are known at present at elevations of 3600–4200 m.

Genus 10. Paracaryum

Hairy and sometimes decumbent perennials. Leaves opposite. Flowers with few or no bracts, small and blue; calyx deeply 5-fid and not enlarged in fruit; corolla funnel shaped, with 5 scales, petals 5 and spreading; stamens 5; ovary 4-lobed; nutlets 4 and depressed.

Three species have so far been found. *Paracaryum glochidiatum* with radical leaves cordate-ovate and long glochidiate nutlets occurs at elevations of 3000–3600 m. *P. heliocarpum* with the radical leaves spatulate-lanceolate and cauline leaves oblong-lanceolate occurs at 3000 m. *P. himalense* is a decumbent hispid plant, with blue flowers and glochidiate nutlets at 3600 m.

Genus 11. Rochelia

Villous or hispid dwarf herbs. Leaves alternate, elongate, subspatulate. Flowers small, blue or white; sepals 5, linear, with auricles beneath basally and enlarging in fruit; corolla cylindrical, with or without scales in throat, petals small, spreading; stamens 5; ovary 2-celled, 2-lobed; stigma capitate; nutlets 2 and single seeded.

Only one species *Rochelia stellulata* has so far been found at 3600–3900 m.

Genus 12. Solenanthus

Hairy perennial herbs. Leaves alternate; radical leaves petiolate. Flowers in corymbs, without bracts; calyx 5-partite and not enlarged in fruit; corolla tubular, with 5 scales in throat, sepals 5 and slightly spreading; anthers small; ovary 4-lobed; nutlets 4.

Solenenthus circinnatus occurs at 3000–3900 m.

Family 25. SCROPHULARIACEAE

About thirty species, belonging to five genera *Lagotis*, *Pedicularis*, *Picrorhiza*, *Scrophularia* and *Veronica* have so far been observed above the timberline in the Northwest Himalaya. Most of these species are found in semi-arid and rocky localities, but a few occur in grassy meadows.

Genus 1. LAGOTIS

Fleshy, decumbent or dwarf, perennial herbs. Leaves largely radical; cauline leaves crenate or incised. Flowers blue or purple, bracteate, in spikes or capitate, without bracteoles, bracts large; calyx spathaceous or with 2 sepals; corolla 2-lipped, the lower lip 2–4-lobed; stamens 2, inserted on corolla tube, anthers large, reniform; ovary 2-celled; stigma capitate or bifid; fruit 2-celled drupe.

Lagotis cachmeriana at 4200 m, *L. decumbens* at 4900–5500 m, *L. glauca* at 3300–4500 m and *L. globosa* at 4200–4900 m are the common species from the Northwest Himalaya.

Genus 2. PEDICULARIS

Generally short, much branched, perennial herbs. Leaves opposite or more usually closely whorled, dentate or pinnatifid. Flowers usually pink, purple or sometimes also yellow with pink, in terminal spikes or in ebracteate racemes; calyx tubular or campanulate, 2–5-toothed; corolla 2-lipped, cylindrical, the upper lip erect, compressed, straight or curved, the lower lip with 3 spreading petals; stamens 4, didynamous; capsule compressed, ovate, lanceolate, falcate or obliquely beaked, with numerous seeds.

There are several species in the Northwest Himalaya. *Pedicularis abratanifolia*, *P. alberti* and *P. bicornuta* are generally found at elevations of about 3000–3900 m; *P. brevifolia* and *P. cheilanthifolia* occur at 3000–4900 m; *P. gracilis* extends from the forest zone upwards to 3900 m; *P. mollis*, *P. pycnantha*, *P. rhinanthoides*, *P. siphonata*, *P. tubiflora* and *P. verticillata* occur at elevations of 3300–4900 m; *P. versicolor* with yellow flowers having pink upper lip is in bloom during June at elevations of 3000–4500 m.

Genus 3. PICRORHIZA

Hairy herbs, with stout perennial rootstock. Leaves subradical, spatulate, serrate, coriaceous. Flowers white, but sometimes also blue, in spikes, bracteate and bracteolate; sepals 5, lanceolate; corolla short, petals 5, limb 2-lipped; stamens 4; ovary 2-celled; capsule ovoid, turgid, septicidal or loculicidal.

One species *Picrorhiza kurroa* occurs at 3000–4500 m.

Genus 4. SCROPHULARIA

Much branched, foetid, perennial, glabrous or more generally puberulous herb, 30–40 cm tall, with rootstock. Leaves opposite, upper leaves alternate, ovate, obovate to oblong, dentate to crenate, glabrous and sometimes coarsely dentate, spatulate or also pinnatifid. Flowers yellow, in thyrsoid cymes or in panicles; sepals 5; corolla tube ovoid or globose, tumid, petals 5, short, flat, 4 uppermost erect, lower spreading; stamens 4, didynamous, with 1 stamen rudimentary; capsule ovoid, septicidal.

The genus is represented by the following five species at high altitudes in the Northwest Himalaya: *Scrophularia calycina* with cordate-ovate and coarsely dentate leaves at 1300–3900 m; *S. dentata* with thick leaves and much branched growth at 3600–4200 m; *S. lucida* and *S. polyantha* at 3300 m and *S. variegata* with spatulate, crenate or pinnatifid leaves at 3000–4200 m.

Genus 5. VERONICA

Much branched, decumbent, usually hoary herbs, about 30 cm tall, with perennial rootstock. Leaves opposite, sessile, oblong, oblong-lanceolate and often serrate. Flowers usually pink coloured, axillary and solitary; calyx 4–5 partite; corolla rotate or tube cylindrical, petals 4–5 and spreading; stamens 2 exserted; capsule bisulcate, often turgid.

Veronica anagallis at 3000–4500 m, *V. capitata* 3000–4500 m, *V. ciliata* at 3300–5000 m and *V. serpyllifolia* with glandular pubescence and creeping branches from the ground at 2500–4000 m represent the typical high altitude forms from the region.

Family 26. LABIATAE

A number of species of this family are typical of cold, wind-blown semi-arid steppes, though some are doubtless mesophytes. Several of them are typical of the elevated regions lying between 4200 and 5000 m. Altogether over two dozen species, contained in *Calamintha*, *Dracocephalum*, *Elsholtzia*, *Hyssopus*, *Marrubium*, *Nepeta*, *Origanum*, *Perovskia*, *Phlomis*, *Scutellaria* and *Thymus* are known at present from the high altitudes of the Northwest Himalaya.

Genus 1. CALAMINTHA

Hairy, procumbent of dwarf herbs, 30–60 cm tall. Leaves ovate, entire or dentate. Flowers violet to pink coloured and in dense axillary whorls; calyx with 13 nerves, 2-lipped, the upper lip 3-fid, lower lip 2-fid; corolla straight, 2-lipped, with the upper lip erect and the lower lip spreading; stamens 4 of which the upper 2 are imperfect; nutlet smooth.

Two species *Calamintha chenopodium* at 1500–3900 m and *C. umbrosa* at 3300–4200 m are known at present (Fig. 31).

Fig. 31. *Calamintha umbrosa.*

Genus 2. DRACOCEPHALUM (Pl. XXII)

Erect or prostrate, glabrous or scaberulous or even woolly plants, about 15–30 cm tall, but often dwarfed, with stout rootstock and fleshy fibres. Radical leaves with long petioles, orbicular-cordate, crenate; cauline leaves sessile; floral leaves orbicular. Flowers generally blue or purple, but sometimes also white; calyx tubular, with 15 nerves and 5 subequal teeth; corolla with erect notched upper lip and 3-fid spreading lower lip with the middle lobe largest.

The high altitude species from the Northwest Himalaya include *Dracocephalum bipinnatum*, *D. diversifolium* (Fig. 32), *D. heterophyllum*, *D. nutans*, *D. speciosum* and *D. stamineum*, occurring generally from 3000 to about 5000 m.

Genus 3. ELSHOLTZIA

Spreading herbs. Leaves opposite. Flowers in simple or panicled, spiked whorls; calyx 5-toothed; corolla tube small, straight or incurved, with oblique limb, 4-fid, the upper lip erect and lower lip spreading; stamens 4; nutlets smooth.

Many species occur at lower elevations in the Himalaya, but *Elsholtzia densa*, *E. eriostachys* and *E. stroblifera* often ascend to elevations of 5000 m.

Genus 4. HYSSOPUS

Low glabrous shrub, much branched from base and mostly not over 30–40 cm tall. Leaves sessile, oblong-linear or lanceolate, entire, obtuse. Flowers blue or pink, in whorls of upto 15, axillary or terminal spikes; calyx with 15 nerves, sepals 5; corolla 2-lipped, the upper lip erect, flat, notched; the lower lip spreading, 3-lobed, the middle lobe largest; stamens 4, exserted, the lower longer; nutlets smooth, triquetrous.

Hyssopus officinalis is in bloom during May–June at 2500–3600 m.

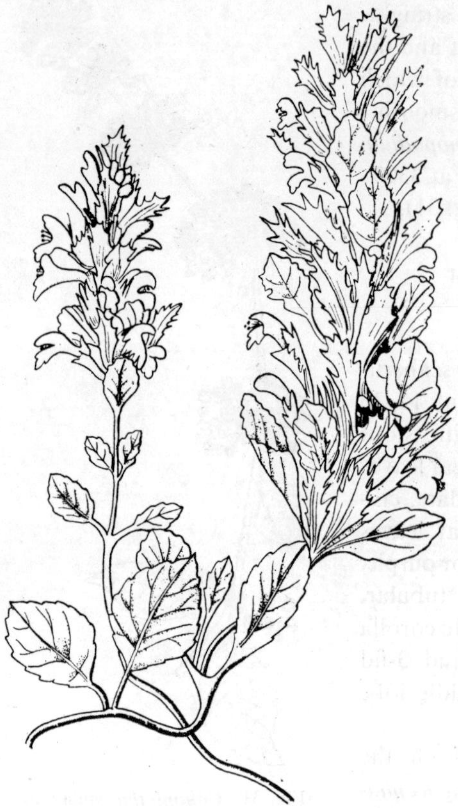

Fig. 32. *Dracocephalum diversifolium.*

Genus 5. MARRUBIUM

Tomentose or woolly dwarf, perennial herbs. Leaves orbicular, ovate, crenate. Flowers with tubular calyx and 5–10-nerved, 5–10-toothed, equal and spinescent; corolla short, annulate, with the upper lip erect and the lower lip spreading.

Marrubium lanatum at 4500–4900 and *M. vulgare* reaching upto 2500 m are sparsely found.

Genus 6. NEPETA

Erect or prostrate, pubescent or woolly, often aromatic, dwarf herbs, with stout rootstock. Leaves opposite, often pinnatifid, entire or crenate. Flowers blue, white or yellow; calyx tubular, with 15 nerves, equally 5-toothed; corolla tube inflated at throat, 2-lipped, the upper lip straight, notched or 2-fid, the lower lip 3-fid with the middle lobe largest; stamens 4, the upper longest; nutlets smooth.

Though many species occur at lower altitudes on the Himalaya, the following few may be considered as characteristic of high altitudes: *Nepeta cataria*, *N. glutinosa*, *N. linearis*, *N. longibracteata*, *N. supina* and *N. thom-*

soni occur up to elevations of 4200–5100 m.

Genus 7. ORIGANUM

Prostrate, stoloniferous or erect villous and dwarf aromatic herbs, generally not exceeding 30 cm in height. Leaves broadly ovate, entire or dentate. Flowers pink, with large bracts; calyx with 10–13 nerves, 5-toothed or 2-lipped; corolla obscurely 2-lipped, with the upper lip notched and 2-fid, the lower lip 3-fid and spreading; flowers often dimorphic, the larger ones bisexual and darker coloured and the smaller ones pistillate and pale pink coloured.

Origanum vulgare is common at 2500–3900 m.

Genus 8. PEROVSKIA

Dwarf, much branched, stellately hairy and aromatic shrub. Leaves opposite, oblong-ovate or lanceolate, crenate. Flowers in small whorls; calyx short and often with cottony hairs, upper lip with 1–3 teeth and small, the lower lip with 2 teeth; corolla 2-lipped, upper lip large and spreading, 3-lobed; stamens 4, with the lower 2 fertile and upper 2 imperfect; nutlets pyriform and smooth.

Perovskia atriplicifolia with dense white stellate hairs occurs generally at elevations of 2200–3900 m.

Genus 9. PHLOMIS

Thick stemmed herbs. Leaves with short or long petioles, broadly ovate-oblong, crenate. Flowers yellow but sometimes pink or white coloured in axillary whorls; bracteoles numerous; calyx 5–10-nerved, often 5-toothed; corolla with concave upper lip

Fig. 33. *Phlomis bracteosa.*

and spreading 3-fid lower lip; stamens 4; nutlets triquetrous.

Phlomis bracteosa (Fig. 33) occurs at 1500–3900 m, *P. macrophylla* at 3000–3900 m, *P. oreophila* at 3900–4200 m and *P. setigera* at 3000–3600 m.

Genus 10. Scutellaria

Prostrate, much branched, often woolly and dwarfed herbs, with yellow flowers; calyx 2-lipped and closed in fruit.

Scutellaria galericulata at 3900 m, *S. heydei* at 4500–4900 m and *S. prostrata* at 2500–4200 m.

Genus 11. Thymus
(Pl. VIII)

Glabrous or hairy, much branched, decumbent or prostrate, aromatic dwarf plant, with woody rootstock. Leaves small, linear, oblong, obovate, entire. Flowers purple but variable in capitate whorls, dimorphic; staminate flowers larger, calyx with 10–13 nerves, 2-lipped; corolla 2-lipped, the upper lip straight, flat, notched; the lower lip 3-fid; sepals ciliate; stamens 4, lower longer; nutlets smooth.

Thymus serphyllum is commonly in bloom during late May to the middle of July at elevations of 2500–4500 m.

Family 27. CHENOPODIACEAE

A small, but widely distributed family of steppes species, often characteristic of rocky localities and ascending upto 4500 m. Some species occur sparsely in grassy meadows. The genus *Chenopodium* ascends often above the timberline and grows as dwarfed and tufted clumps at 3900–4200 m. *Atriplex* is also sometimes found at 3600–4200 m and *Haloxylon* as a much branched and hairy shrub at 3000–4200 m. The typical high altitude species belong to *Atriplex*, *Axyris*, *Chenolea*, *Corispermum*, *Eurotia*, *Kochia*, *Microgynoecium* and *Salsola*.

Genus 1. Axyris

Stellately hairy, dwarf plants, about 15 cm tall. Leaves alternate, small, entire, about 10–35 mm long, generally pale yellow or reddish-brown, with obscure veins, short petiole, ovate or oblong, obtuse or acute. Flowers minute, monoecious, without bracts; staminate flowers in terminal clusters, with 3–5 hyaline sepals and 2–5 stamens; pistillate flowers axillary, solitary or mingled with the staminate flowers, sepals 3–4, unequal; utricle membranous, embraced by the enlarged sepals, obovoid and slightly compressed, winged or with two auricles at the tip.

Axyris amaranthoides is common on grassy meadows at elevations of 2500–4200 m.

Genus 2. CHENOLEA

Silky, villous or tomentose, much branched dwarf shrubs, about 10–30 cm tall. Leaves alternate, linear. Flowers minute, bisexual; pistillate flowers woolly without bracts; perianth 5, incurved; stamens 5; utricle membranous, included in the coriaceous perianth with 5 spines.

Chenolea divaricata occurs at 3300–4200 m.

Fig. 34. *Eurotia ceratoides*.

Genus 3. Corispermum

Diffusely branched from the ground and about 15–25 cm tall. Leaves alternate, linear-ovate, obtuse with one vein; floral leaves margined white. Flowers bisexual in spikes; sepals 1–3 and unequal, scarious; bracts green with white margin; stamens 1–5 and unequal; utricle winged.

Corispermum hyssopifolium occurs at 3000–4500 m.

Genus 4. Eurotia

Hoary or woolly or also tomentose and often densely much branched herb or dwarf shrubs, about 60–120 cm tall or more generally much shorter. Leaves alternate, small, 6–30 mm long, almost obsolete petiole or with very short petiole, linear-oblong or lanceolate, obtuse, rarely ovate, with recurved margin. Flowers minute, unisexual; staminate flowers spicate, ebracteate; sepals 4, obovate, stamens 4 with filiform filament; pistillate flowers axillary, with two coriaceous bracts which may be connate and 2-beaked and clothed with silky villous hairs and reddish coloured; perianth absent; utricle ellipsoid, compressed, membranous.

Eurotia ceratoides occurs from 2500 to 4500 m (Fig. 34).

Genus 5. Kochia

Villous, much branched tufts of dwarf perennial shrub, with stout rootstock. Leaves alternate, sessile, linear, entire. Flowers minute; pistillate flowers axillary, ebracteate, perianth with 5 incurved lobes enclosing the utricle and with 5 spreading wings; stamens 5; utricle membranous and depressed.

Kochia prostrata occurs at 3000–4200 m.

Genus 6. Microgynoecium

Diffuse, prostrate or erect dwarf plants with slender branches 2 cm long, rising from the ground. Leaves fleshy, alternate, petiolate, entire. Flowers small, clustered, monoecious and concealed by the leaves, staminate flowers without bracts, perianth 5-lobed and hyaline, stamens 1–4 with long filaments; pistillate flowers with 2 bracts and without perianth; utricle hyaline, obliquely ovate, compressed, with 1–2 auricles at the tip.

Microgynoecium tibeticum occurs at 3300–4500 m.

Genus 7. Salsola

Spinescent herbs. Leaves alternate, sessile, short, ovate or linear.

Flowers rose coloured, bisexual, sepals 4–5 with spreading wings enlarging in fruit and embracing the utricle; stamens 5 or fewer; utricle membranous or fleshy, globose.

Salsola kali ascends to elevations of 3300–4500 m.

Family 28. POLYGONACEAE

Twenty-six species of this family are known at present in the high altitude areas of the Northwest Himalaya. Most of them are characteristic of arid steppes, but some few occur in mesophile grassy meadows.

Genus 1. OXYRIA

Glabrous, erect, tufted and succulent dwarf herbs, with stout and tufted rootstock. Leaves radical, orbicular-cordate, reniform. Flowers bisexual, sepals 4, the outer two ovate and reflexed; the inner two short; nut biconvex and with two wings.

Oxyria digyna is common at 3600–5500 m.

Genus 2. POLYGONUM
(Pl. X, XX)

Mostly erect herbs, sometimes stunted and prostrate or spreading as much branched plants on the ground. Leaves alternate or opposite, usually entire, with tubular membranous stipule. Flowers bisexual in axillary or terminal clusters, with membranous bracts and bracteoles. Perianth 4–5 or 3-cleft; stamens 5–8; ovary compressed or trigonous; styles 2–3 with capillate stigma; nut included in the enlarged perianth.

Numerous species occur in the forest and at higher elevations, especially in marshy and grassy meadow areas. The high altitude species from the Northwest Himalaya include the following: *Polygonum acaule* at 5100 m, *P. affine* at 2800–4200 m, *P. cognatum* at 3000–5000 m, *P. convolvulus* at 3300–5000 m, *P. delicatulum* at 3000–4900 m, *P. filicaule* at 3000–4900 m, *P. glaciale* at 2500–3600 m, *P. islandicum* at 3300–4200 m, *P. molliaeforme* at 4200–5000 m, *P. nummularifolium* at 3900–4500 m, *P. persicaria* at 3000–4200 m, *P. paronychioides* at 2800–3600 m, *P. polystachyum* at 2500–4900 m, *P. rumicifolium* at 3000–4200 m, *P. sibiricum* at 3300–5000 m, *P. sphaerostachys* at 3000–4500 m, *P. tuberosum* at 3300–5000 m, *P. tubulosum* at 2500–3600 m, *P. vaccinifolium* at 3000–4200 m and *P. viviparum* at 3000–4500 m.

Genus 3. RHEUM

Stemless dwarf plants. Leaves coriaceous, entire or dentate or even lobed, with scarious stipules. Flowers densely clustered together and usually bisexual, sepals 5, stamens 6–9, ovary 2–4 angled, nuts 2–4-winged.

Rheum emodi occurs at 3000–3900 m, *R. spiciforme* at 3000–5000 m and *R. webbianum* at 3000–4200 m.

Genus 4. Rumex

Often prostrate or stunted plants with ovate or oblong leaves. Flowers hermaphroditic, monoecious or dioecious in axillary cluster whorls in simple or racemose panicles; perianth cleft into 6, with the inner 3 sepals acrescent and developing into conspicuous red or rose coloured wings on the seeds; stamens 6; stigma fimbriate.

Rumex acetosa is in bloom from May through to July at elevations from 2500 to 3900 m.

Family 29. ELAEAGNACEAE

Genus 1. Hippophaë

Spinescent and often curiously stunted and densely branched shrub. Leaves alternate, linear to obovate, with silvery circular scales; twigs also covered by the same scales. Flowers dioecious and appearing on the older parts of the plant; staminate flowers appear in axillary clusters with deciduous bracts and perianth two opposite; fruit yellow or scarlet, globose membranous utricle included within the succulent perianth.

The commonest and most widely distributed species *Hippophaë rhamnoides* is very characteristic of arid and wind blown steppes.

Family 30. THYMELEACEAE

Genus 1. Stellera

Dwarf herbs, hardly 20 cm high, with stout rootstock. Leaves alternate or opposite, glabrous, sessile, elliptic-lanceolate, acute to acuminate, about 20 mm long. Flowers yellow and sweet scented in terminal involucrate heads; perianth tube cylindrical, with 4 short lobes, glabrous and spreading; stamens 8, biseriate and sometimes also 10; ovary sessile and 1-celled; stigma capitate; fruit dry and enclosed in perianth base.

An unidentified species with dwarfed growth was found at 2800–3600 m during June–July.

Family 31. EUPHORBIACEAE

This family is represented by the following species of *Euphorbia* growing at high altitudes in the Northwest Himalaya: *Euphorbia himalayensis* with coriaceous leaves at 3000–3900 m; *E. kanarica* at 4500 m; *E. stracheyi* at 3600–4500 m; *E. thomsoniana* at 3000–3600 m; *E. tibetica* at 3000–4500 m and *E. wallichii* at 3000–3600 m. Many other species like *E. heliopa* occur at lower elevations and often serve as source for sparse and ephemeral pockets growing at higher elevations above the timberline, where the plants either do not bloom or are sterile and greatly stunted. The family is apparently a very recent intrusive element of the high altitude ecosystem.

3. MONOCOTYLEDONS

Family 32. HAEMODORACEAE

Genus 1. Aletris

Perennial herbs. Leaves radical, tomentose, narrow, 5–7-nerved, linear and resembling the blade of a grass about 7–20 cm long; flowers green, small in spikes of leafless bracteate scape, with perianth tube short and subcampanulate; stamens 6, with short filament and ovate anthers; capsule 3-valved loculicidal.

Aletris nepalensis is sparse at elevations of 3000–4500 m.

Family 33. IRIDEAE

Genus 1. Iris

Perennial herbs with bulbous or creeping rootstock. Leaves narrow, ensiform. Flowers generally yellow or often lilac coloured; perianth with large outer segments reflexed and the inner smaller segments suberect; stamens inserted at base of the outer segments; style stout; stigma petaloid; capsule coriaceous and with 3–6 ribs.

Iris hookeriana is a dwarf plant with linear leaves; the outer lobes of perianth obovate, cuneate, purple-blue and blotched darker, claw white with violet veins, hairs of beard white, yellow ripped, commonly in bloom during May through to July at elevations of 3600 m.

Family 34. LILIACEAE

The members of this family are important elements of grassy meadows. Decidedly gregarious and nearly pure stands or also associations with Ranunculaceae and Umbelliferae often cover extensive areas, almost up to the glacier edge. The plants are generally in bloom during mid summer.

The common species belong to *Allium*, *Eremurus*, *Fritillaria*, *Gagea*, *Lloydia*, *Theropogon*, *Trillium* and *Polygonatum*. *Allium* is the dominant genus.

Genus 1. Allium
(Pl. XVIII)

Scapigerous herbs with coated bulbs. Leaves linear. Flowers white, yellow or red or purple, in capitate umbels, enclosed in 1–3 membranous spathes and stellate; sepals 6, free; stamens hypogynous or inserted on the perianth; ovary trigonous and 3-celled.

Fifteen species have so far been found in grassy meadows at elevations above 3300 m, forming associations with *Caltha*, *Trollius*, etc.

Allium atropurpureum with its bulb not on the rootstock, leaves flat, spathe shorter than head; leaves linear-oblong, shorter than the fistular scape; flowers rose-purple, stellate. Blooms during June at 2500–3000 m. *A. atrosanguineum* occurs at somewhat higher elevations and is common enough at 3600 m. *A. blandum* has flat leaves, about 7–25 cm long, with the bulbs clustered on the rootstock, flowers purple or pale pink, globose head and densely flowered, with sepals oblong and in bloom during May–June at 3900–5100 m. *A. caeruleum* is in bloom at 3900 m during June. *A. fedschenkoanum* was found at 3600 m. *A. govanianum* is perhaps the commonest species on grassy meadows in the Northwest Himalaya and is conspicuous for its numerous leaves, the basal subdistichous, linear, flat, obtuse and subequal to the angled scape; heads many flowered, the flowers being white coloured, stellate, stamens shorter than perianth; the plant is in bloom during June–July at elevations of 3600–3900 m. *A. jacquemontii* with bulbs tufted cylindrical, basal leaves filiform; heads globose; flowers lilac and stamens longer than the perianth; common at 3600–4200 m. *A. loratum* with flat leaves longer than the cylindrical scape and white campanulate flowers in bloom during June–July at 3000–4200 m. *A. odorum* with numerous leaves, which are narrow, linear, flat, keeled and shorter than the scape; flowers small, white; stamens shorter than the perianth; blooming during June–July at elevations of 3000–4200 m. *A. oreoprasum* has linear, flat leaves, shorter than the scape; flowers in hemispherical heads, rose coloured and with the stamens shorter than the perianth, found in bloom during June to July at elevations of 3000–5100 m. *A. pallasii* and *A. platyspathum* are generally confined to the inner and more arid localities. *A. schrenkii* has flat, linear leaves, shorter than the scape, which is striated; heads globose; flowers purple and campanulate; stamens longer than the perianth; generally in bloom during June at elevations of 3600–4200 m. *A. thomsoni* has somewhat stout leaves and red-purple flowers at 3600 m. *A. victorialis* is recognized by its petioled leaves, elliptic to oblong-lanceolate and greenish-white or yellow flowers at elevations of 3000–3600 m.

Genus 2. EREMURUS

Perennial herbs. Leaves radical, linear, glabrous, 2–7 cm long; scape stout, 60–100 cm long; flowers 2 cm large, white, with subulate-lanceolate bract; capsule rugose.

Eremurus himalaicus is commonly in bloom from June to August at elevations of 2500–4000 m.

Genus 3. FRITILLARIA
(Pl. XVII)

Plants generally resembling *Lilium*, but with campanulate perianth, the segments of which may be spreading; stigma 3-fid; flowers solitary; leaf elliptic-lanceolate.

Two species *Fritillaria oxypetala* and *F. roylei* with yellow flowers and whorled leaves, are generally in bloom on grassy meadows at elevations of 3000–3900 m during May–June.

Genus 4. GAGEA

Bulbiferous herb usually not taller than 5–8 cm. Leaves single, linear or lanceolate; scape with umbelliform leafy cymes or corymbs of yellow stellate flowers; capsule membranous.

Gagea lutea is in bloom in grassy meadows during April–June at elevations ranging from 2500 to 3900 m.

Genus 5. LLOYDIA

Dwarf bulbous herb. Leaves filiform, wiry, 5–20 cm long; flowers small, white or yellow, solitary, with perianth funnel shaped, with 6 segments and 3–5 nerved; stamens 6.

Lloydia serotina occurs in grassy meadows at 3600–5100 m.

Genus 6. TRILLIUM

Herbs with creeping rootstock, simple erect stem. Leaves 3-whorled, 3–5-nerved, reticulate, shortly petiolate, ovate or ovate-cordate; flowers solitary, sessile or pedicelled, perianth persistent, sepals 6, free, spreading; fruit a berry.

Trillium govanianum occurs in grassy meadows at 2500–3600 m.

Other genera

Theropogon pallidus is a grasslike scapigerous plant with white flowers that is sometimes found growing immediately above the timberline in the Northwest Himalaya. *Polygonatum verticillatum*, with lilac flowers, occurs sparsely at elevations of 3000–3600 m.

Family 35. JUNCACEAE

Genus 1. Juncus

Erect, tufted or densely clustered herbs, often with creeping rootstock. Leaves flat or cylindrical; flowers usually terminal, bisexual, green, white, hyaline, coriaceous; perianth in two series; stamens 6.

About half a dozen species have so far been collected, mostly from semi-arid and rocky localities, and sometimes also from grassy meadows. The plants often form pure stands in many areas or occur in association with grass.

The common species from the Northwest Himalaya include *Juncus himalensis* ascending up to 4200 m, *J. leucomelas* characteristic of elevations of 3600–4500 m, *J. membranaceus* ascending almost to 3900 m, *J. sphacelatus* also reaching to 3900 m and *J. triglumis* occurring at 3600–4500 m.

Family 36. AROIDEAE

Genus 1. Arisaema

Tuberous herbs. Leaves trisect or pedatisect, with the leaflets whorled. Spathe deciduous, tube convolute, limb sometimes tailed; spadix included, with appendages excessively long and filiform; staminate flowers many, sessile, anther 2–5, oblong; pistillate flowers densely crowded; ovary 1-celled; style short or absent with the stigma disciform; spadix yellow-pink, with the tail about 30–40 cm long and dark purple. One or two unidentified species generally grow at elevations of 2500–3000 m, but often ascend to 4200 m in the Northwest Himalaya, especially in hygrophile or mesophile localities. The genus and the family appear to be a comparatively recent intrusion in the high altitude ecosystem and apparently pushing upwards at present.

Family 37. CYPERACEAE

Genus 1. Carex

The plants of this genus are generally very much like *Cobresia* in appearance and in distribution. The high altitude species from the Northwest Himalaya include *Carex alpina* ascending to 4500 m, *C. cruenta* also found as high as 4500 m, *C. haematostoma* extending from the upper limits of the forest to 5100 m, *C. incurva* occurring at 3000–3900 m, *C. linearis* occurring at 3300–4200 m, *C. moorcraftii* typical of elevations 4500–5100 m, *C. nivalis* occurring at 3000–5100 m, *C. stenocarpa* (Fig. 35) at 3000–3900 m, *C. tristis* at 3900–4500 m, *C. ustulata* at 3600–5100 m and *C. vulgaris* found at 3000–3900 m.

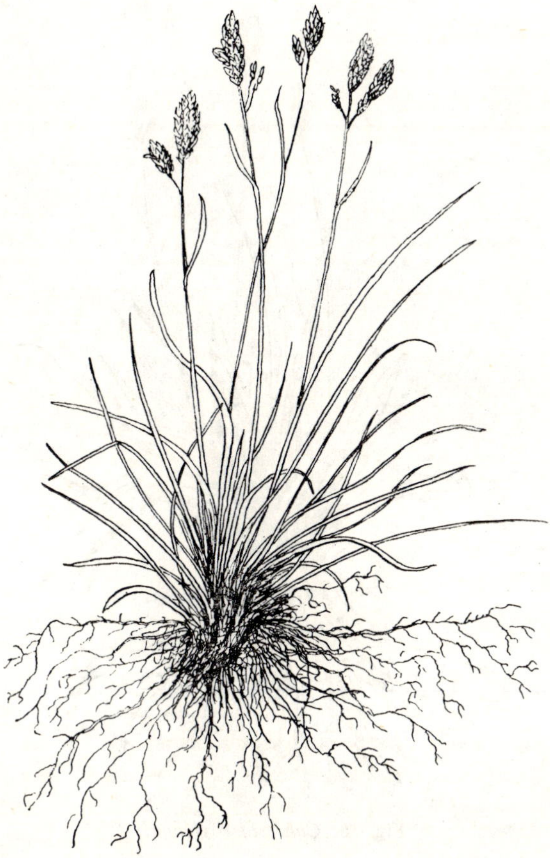

Fig. 35. *Carex stenocarpa.*

Genus 2. Cobresia

Glabrous perennials with simple stem. Leaves near base; spike cylindrical or oblong, terminal; spikelets staminate or pistillate; style 3-fid.

Several species like *Cobresia filicina* at 3000 m, *C. nitens* at 3600–4500 m, *C. pygmaea* at 3600–4500 m, *C. royleana* (Fig. 36) at 3000–5000 m, *C. schenoides* at 3000–5000 m and *C. stenocarpa* are characteristic of arid steppes localities. They form members of character associations.

Genus 3. Scirpus

Short and stout stems, with leaves often only basal from stolons and rhizomes; umbels often compound, spikelets many flowered; glumes 1–2, the lower glume empty, followed by fertile flowers; stamens 3–1; style short or long; nut trigonous or subdiscoid.

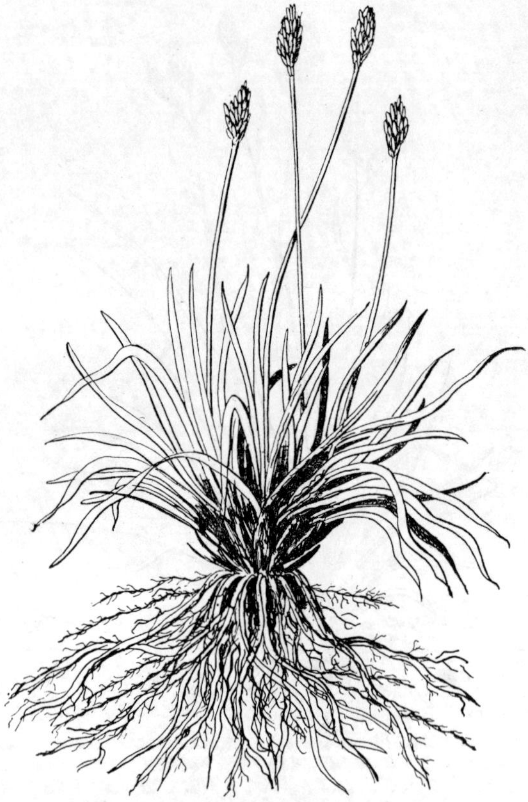

Fig. 36. *Cobresia royleana.*

The only species found at elevations of 3600–4200 m in the Northwest Himalaya is *Scirpus pumilus*, which has a wide distribution in the north (*vide infra*). The plant is not very common in most localities and often occurs in curiously isolated pockets in moist areas.

Family 38. GRAMINEAE

Genus 1. Bromus

Erect grass, with narrow flat leaves. Spikelets multifid, laterally comp-pressed, the uppermost imperfect; glumes many, the first two glumes empty, unequal, 1–5-nerved, persistent; flowering glumes oblong; palea bifid, ciliate, keeled; lodicules 2, oblanceolate.

Bromus crinitus occurs at 3000–4200 m, *B. inermis* at 3000–3900 m, *B. macrostachys* at 3600 m, *B. oxyodon* at 3000–3600 m and *B. patulus* ascends up to 4200 m.

Genus 2. Calamagrostis

Stout or slender and narrow leaves; spikelets in panicles; glumes 3; first and second glumes equal or unequal, lanceolate; third glume shorter, hyaline; palea may be absent.

Calamagrostis epigejos occurs at 2500–4500 m and *C. littorea* ascends sometimes to 3000 m.

Genus 3. Danthonia

Densely tufted grass; spikelets 2-fid, in panicles, rachilla articulate, hairy; glumes 5 or more; the first and second glumes empty, subequal, keeled, persistent; flowering glumes ciliate, with 7–9 nerves; palea broad; lodicules 2, fleshy; stamens 3.

Danthonia cachemyriana is common at 3000–4500 m and *D. exilis* occurs at 4500 m.

Genus 4. Deschampsia

Grass of swampy localities, with flat or convolute leaves; spikelets 2-fid; rachilla articulate; glumes 4, the first two glumes empty, subequal or sometimes the second glume is somewhat larger and persistent, third and fourth glumes subhyaline; lodicules ovate, fleshy.

Deschampsia caespitosa and *D. koelerioides* occur at 3000–5000 m.

Genus 5. Deyeuxia

Grasses with the general appearance of *Calamagrostis*, but with the rachilla of spikes produced beyond the third glume, which is also glabrous. *Deyeuxia scabrescens* occurs at 3000–4900 m.

Genus 6. Elymus

Tufted grass. Leaves flat or convolute. Spikelets 3-fid or multifid, sessile; glumes one and two empty, unequal, persistent; flowering glume rigid, 5–7-nerved and converging to tip; palea keeled, ciliate; lodicules entire or ciliate. *Elymus sibiricus* is a densely tufted grass widely distributed at elevations of 3000–5500 m.

Genus 7. Festuca

Tufted grasses. Leaves flat or convolute or setaceous. Spikelets 3-fid or multifid in panicles; glumes 5 or more; first two glumes subequal, enlarging, costate, persistent; first glume 1–3-nerved; second glume 3–5-nerved; flowering glumes 5–7-nerved; palea 2-costate; lodicules 2.

Several species occur at high altitudes in the Northwest Himalaya. *Festuca altaica* is common at 3300–4500 m, *F. kashmirica* at 2500–4200 m, *F. nitidula* at 3600–4500 m, *F. sibirica* at 3600–4200 m and *F. valeriaca* at 3000–5000.

Genus 8. Glyceria

Tufted grass. Leaves flat, narrow, sometimes also convolute. Spikelets multifid; first and second glumes hyaline, empty, persistent, 1–5-nerved; flowering glumes oblong or broader, 3–9-nerved; lodicules 2. An unidentified species is fairly common in semi-arid and rocky localities at elevations of 3000–5000 m.

Genus 9. Poa

Leaves flat or convolute. Spikelets 2–6-fid, lax or in close panicles, compressed; glumes 4; first two glumes empty, persistent, 1–3-nerved, keeled; flowering glumes 5–7-nerved; palea shorter than glumes, 2-keeled; lodicules 3.

About half a dozen species occur up to 5100 m: *Poa alpina*, *P. attenuata*, *P. flexuosa*, *P. nemoralis*, *P. pratensis* and *P. tibetica*.

Genus 10. Stipa

Tufted perennial grass. Leaves generally convolute. Spikelets in loose panicles, sometimes dorsally compressed; glumes 3; the first two glumes empty; third glume rigid; palea 2-nerved; lodicules 3 and large.

Several species occur at high altitudes in the Northwest Himalaya: *Stipa basiplumosa* at 4500–5100 m, *S. breviflora* at 4500 m, *S. capillata* at 2800–3300 m, *S. hookeri* at 3000–5000 m, *S. mongolica* at 4200–4900 m, *S. orientalis* and *S. purpureum* at 4800–5100 m and *S. sibirica* at 3000 m.

4. SUMMARY

This chapter provides the necessary basis for discussions on distribution, phytogeography and origin of the high altitude vegetation of the Himalaya.

A taxonomic treatment is not aimed at in this chapter, but the brief notes on about two hundred genera of high altitude plants serve to illustrate their ecological specializations. The genera are arranged alphabetically under their respective families, with a brief list of common or otherwise interesting species. The plants listed here do not exhaust the high altitude flora, but include only the few that have been actually collected or observed by the author. Future explorations are bound to bring to light many more plants. The dominant families include Cruciferae, Caryophyllaceae, Compositae, Primulaceae, Cyperaceae and Gramineae.

A number of species, occurring at high altitudes in the Northwest Himalaya, are either identical with or are very close to those recorded from the Pamirs, Central Tien Shan and often also the Altai Mountains. While the flora of the Pamirs may be said to be fairly adequately known, thanks to the pioneer labours of eminent Russian botanists like Korovin, the flora of the high altitudes of the Himalaya is only very imperfectly known at present. There is urgent need not only for intense botanical explorations, but also for careful taxonomic revisions of the Himalayan high altitude plants.

Chapter V

Vertical Distribution

1. INTRODUCTION

ALTHOUGH there is extensive literature on the altitudinal zonations of organisms on mountains [4, 13, 33, 75–78, 105, 106], very little is known of the vertical distribution of high altitude plants [104, 117]. In the Northwest Himalaya, the vertical distribution of high altitude plants presents a number of peculiarities. In striking contrast to the other parts of the Himalaya, in the Northwest Himalaya the vertical distribution is profoundly influenced by factors, which are peculiar to the region. The general picture of distribution is also often complicated by the physical continuity and phytogeographical unity with the Pamirs. The vertical distribution in the region cannot, strictly speaking, be completely isolated from horizontal distribution. Enormous distances and extensive elevated areas of a series of mountain ranges are involved, so that we have here instead of simple vertical distribution, a complex of altitude-latitude, floristic-vegetation succession. This fact is reflected both in the general physiognomy and in the floristic composition at different altitudes. The pattern is further complicated by the southeast-northwest trendlines of the mountain ranges and by the south-north slope differences. The north slopes are generally less steep and more extensive than the south slopes of the same range. The influence of altitude on distribution is thus often telescoped, obscured or even completely obliterated or inverted. The occurrence of a plant in a given altitudinal belt, especially towards the western end, or its absence in this particular belt or at lower or higher elevations in the eastern end, is not necessarily part of the altitudinal distribution pattern, but basically one of horizontal distribution in the Northwest Himalaya. Strictly speaking, we need a tridimensional treatment [116] for a satisfactory representation of the distribution of the high altitude plants of the Northwest Himalaya. An attempt is made in the following pages to give a simplified account of the salient facts of the vertical distribution of the high altitude plants.

2. THE UPPER ALTITUDINAL LIMITS OF HIGH ALTITUDE PLANTS

The massive and abrupt fall in abundance of genera and species at the timberline and the progressive thinning out with increase in altitude, observed in other organisms, especially insects, do not hold good for the high altitude plants [75]. The fall in their abundance with altitude is far more gradual and becomes conspicuous only at elevations close to the permanent snowline. We find, therefore, that almost up to elevations of 4500–5000 m there is scarcely any loss either in the diversity or in the abundance of individuals. While no doubt many genera and species, which are charactersitic of the upper reaches of the forest, naturally tend to disappear above the timberline, there is a very pronounced replacement by perhaps even larger numbers of genera and species that are increasingly characteristic of the higher elevations. Further, many of the subalpine-zone plants tend to be secondarily vegetatively modified for the prevailing conditions, which do not seem to exclude them altogether from colonizing the higher elevations. This leads to many peculiarities in the altitudinal distribution pattern. While most high altitude plants seem unable to flourish below the timberline, there does not seem to be any significant effect on the altitudinal limits higher above.

The highest altitude record for an Angiosperm in the world is 6300 m, at which elevation grows the Crucifer *Christolea hymalayensis* on Mt. Kamet in the Western Himalaya [77, 101]. Cushions of *Stellaria decumbens* were observed by Swan [111, 112] at elevations of 6140 m on Mt. Makalu in the Central Himalaya. Other Angiosperms found growing up to 6000 m in the Himalaya include *Allardia glabra*, *Parrya lanuginosa*, *Pegophyton scapiflorum*, *Gentiana urnula*, *Selinum corticoides*, *Delphinium brunonianum*, etc. Many more species are found at elevations between 5500 and 6000 m than at extreme elevations in the Northwest Himalaya. The Gymnosperm *Ephedra* is represented by several xerophytic or semi-xerophytic species in the Northwest Himalaya and Pamirs up to elevations of 5200 m. As already indicated, moss and lichens naturally grow at much higher elevations than the Angiosperms and may be found abundantly at 6300–6600 m. Yeast and fungi have been reported at elevations of 7600–8400 m on Mt. Everest [112]. The upper limits of altitude of plant growth are generally lower in the Northwest Himalaya than in Sikkim or Nepal Himalaya. Even within the area of the Northwest Himalaya there is a very pronounced fall westward in the value of the upper altitudinal limit. This peculiarity is particularly observed in genera and species, which are distributed from the eastern to the western extremity of the Himalaya. *Cochlearia* occurs, for example, at elevations up to 5200 m in Sikkim-Nepal Himalaya, but does not ascend over 5000 m in the Northwest Himalaya. The upper altitude limit of *Lychnis* is 5900 m in the east, but only 5500 m in the Northwest Himalaya. *Allardia* occurs only up to 5500 m in the Northwest Himalaya, but has been reported by Swan [111] at 6000 m in the Central Himalaya. *Tanacetum* occurs at 4200 m in the Northwest and 4800 m in the east. The limits for *Senecio* are 4500 m and 4900 m respectively, for *Androsace* 5000 m and 5200 m, *Gentiana* 4800 m and 6000 m. *Saxifraga engleriana* occurs up to 5730 m on Mt. Makalu, but in the Northwest Himalaya no species of *Saxifraga* has so far been found above 5500 m. Most mesophytes and hygrophiles ascend higher in the east than in the west, but xerophytes occur higher in the west.

In addition to the east-west gradient, there is also a perceptible north-south transition in the upper limits of altitude. In the Northwest Himalaya, as we pass northward, the mean upper altitudinal limit rises at first and after reaching a maximum altitude of about 6000 m on the Great Himalaya,

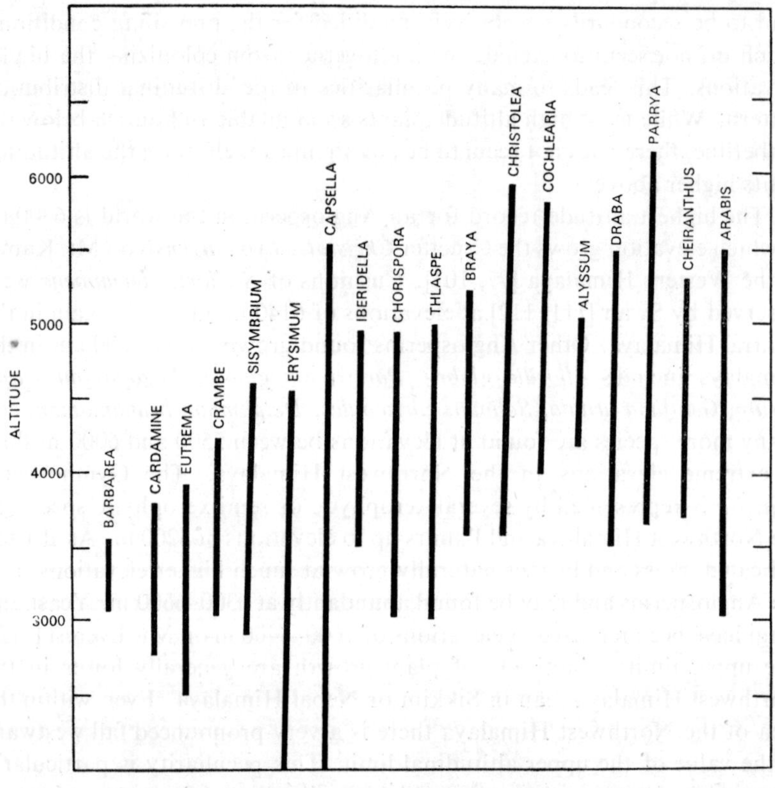

Fig. 37. Histograms showing the fluctuations in the altitudinal ranges of high altitude genera of Cruciferae in the Northwest Himalaya. Each bar merely indicates the lowest and the highest altitudes at which a high altitude species of the genus has been collected so far.

dips again almost to 4900 m. Interesting genera like *Oxytropis* and *Sibbaldia* reach their maximum altitude of 6000 m in the Pamirs, but do not ascend above 5500 m in the Northwest Himalaya. Most other genera, which are common to these two regions (*vide* Chapter VI), are generally stabilized at elevations of about 4900–5000 m, both in the Pamirs and the Northwest Himalaya. It is interesting to observe that generally speaking the upper limits are higher in heavily glaciated areas or at least in areas which were formerly glaciated than in the scantily glaciated or non-glaciated parts. The limits are somewhat higher in certain families like Cruciferaceae, Caryophyllaceae, Gramineae, etc., than in others. As a whole, the mean upper limits of most Angiosperms appear to lie within a narrow belt oscillating about 5500 m. The reader will find an interesting discussion on the altitudinal limits of vascular plants in Webster [122] and Zimmermann [123].

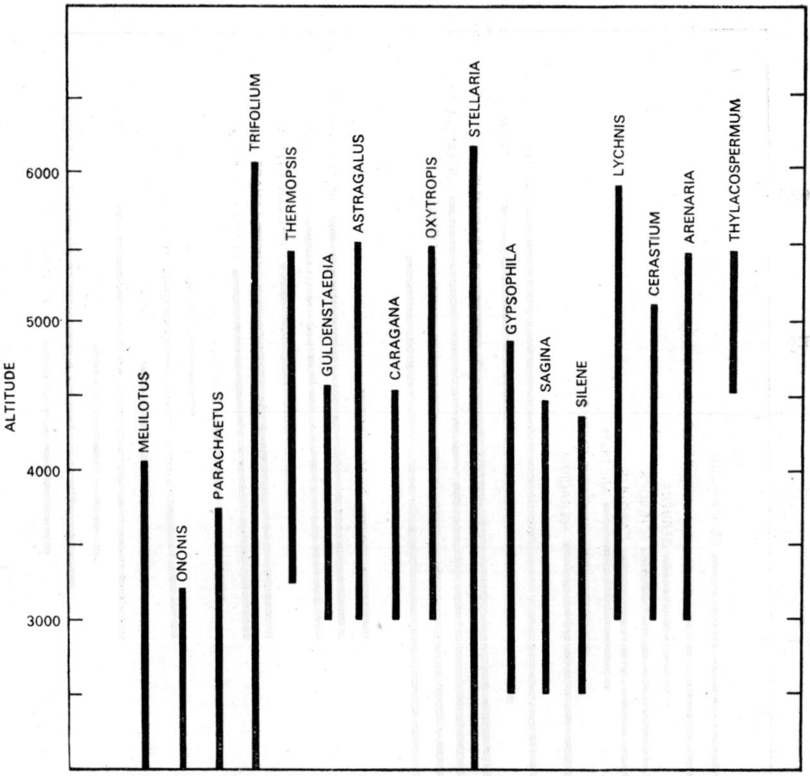

Fig. 38. Histograms showing the fluctuations in the altitudinal ranges of high altitude genera of Leguminosae and some other families in the Northwest Himalaya.

3. ALTITUDINAL RANGE

The altitudinal range of a species is defined conveniently by the upper and lower limits of its occurrence, irrespective of whether the plant is in bloom or sets seeds at these limits. The range may be wide or narrow, depending upon whether these limits are high or low. Though the lower limit is theoretically the timberline, many plants often descend well below it. Similarly, as already pointed out, the upper limit is also subject to considerable and indeed great variation. The altitudinal range thus differs in different families and even genera and is also different in different ecological types, such as mesophytes or xerophytes. It must be clearly understood here that altitude, as such, has very little meaning, unless it is also indicated at the same time, to which mountain range or slope aspect the altitude record refers. In the following table are summarized the mean altitudinal

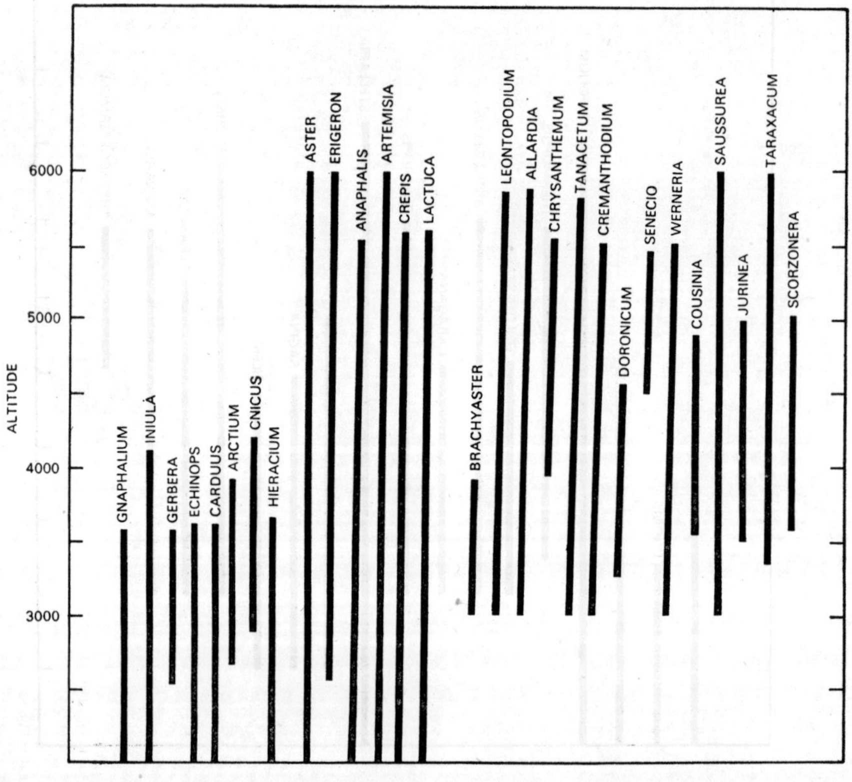

Fig. 39. Histograms showing the fluctuations in the altitudinal ranges of high altitude compositae in the Northwest Himalaya.

Fig. 40. Histograms showing the fluctuations in the altitudinal ranges of other families of Phanerogams in the Northwest Himalaya.

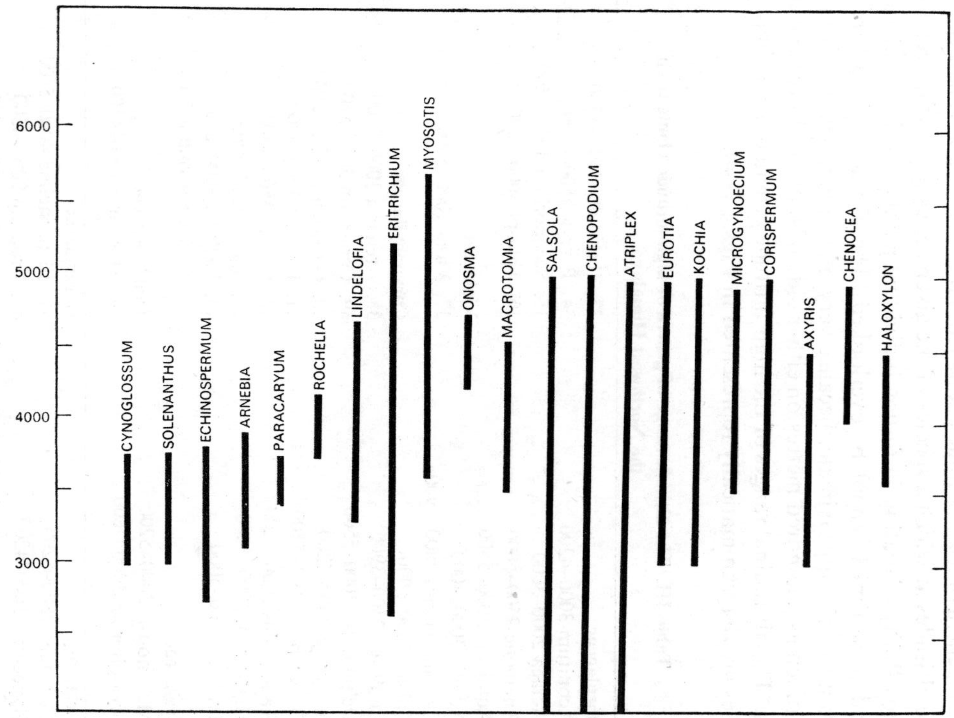

range of genera of high altitude plants from the Northwest Himalaya, based on the altitudes at which a species of the given genus has so far been found. The possibility of finding in future, individual plants growing at elevations beyond these limits cannot be overlooked. The altitudes given here must be taken as only approximate, because very often plants may be found up to as much as a hundred metres on either side of both the lower and upper limits. The altitudinal ranges of the high altitude genera of some important families are diagrammatically represented in Figs. 37–40.

Table III. The altitude range of genera of high altitude plants from the Northwest Himalaya

Ranunculaceae
1. Aconitum 3000–4200
2. Actaea 2000–3000
3. Anemone 2500–5000
4. Aquilegia 3000–5500
5. Adonis 3000–4000
6. Callianthemum 2800–3900
7. Caltha 3000–5000
8. Cimifuga 2500–3900
9. Delphinium 3000–5500
10. Isopyrum 3000–5200
11. Oxygraphis 3600–5500
12. Ranunculus 2000–5200
13. Thalictrum 3000–5200
14. Trollius 3300–5000

Papaveraceae
15. Meconopsis 3000–5200
16. Stylophorum 3000–3600

Fumariaceae
17. Corydalis 2800–5200
18. Hypecoum 3000–4200

Cruciferae
19. Alyssum 4200–5200
20. Arabis 3000–5000
21. Barbarea 3000–3300
22. Braya 3600–5500
23. Capsella 3000–5900
24. Cardamine 2500–3600
25. Cheiranthus 3600–5200
26. Chorispora 3000–5500
27. Christolea 3600–6000
28. Cochlearia 4000–6000
29. Crambe 3000–4000
30. Draba 3600–5500
31. Erysimum 1500–4500
32. Eutrema 1800–3900
33. Iberidella 3600–5500
34. Parrya 4500–6100
35. Sisymbrium 3000–4500
36. Thlaspi 3000–5200

Violaceae
37. Viola 1000–5000

Caryophyllaceae
38. Arenaria 3000–5500
39. Cerastium 3000–5100
40. Gypsophila 2500–3900
41. Lychnis 3000–5900
42. Sagina 2500–4500
43. Silene 2500–3900
44. Stellaria 3000–6200
45. Thylacospermum 4500–5500

Tamariscaceae
46. Myricaria 3000–4500

Geraniaceae
47. Biebersteinia 4000–5500
48. Erodium 3000–4500
49. Geranium 2500–4500
50. Oxalis 2500–3600

Leguminosae
51. Astragalus 3000–5500
52. Caragana 3000–5500
53. Guldenstaedtia 3300–4500
54. Melilotus up to 4000
55. Onobrychis 3900
56. Ononis 1500–3000
57. Oxytropis 3000–5500
58. Parochaetus 1800–3600
59. Thermopsis 3300–5500
60. Trifolium 1800–6000

Rosaceae
61. Fragaria 1600–3900
62. Geum 1800–3300
63. Potentilla 1800–5500
64. Spiraea 1300–3600

Saxifragaceae
65. Parnassia 3600–4500
66. Saxifraga 2800–5500

Crassulaceae
67. Sedum 2800–5500
68. Sempervivum 3000–4500

Onagraceae
69. Epilobium 2800–4800

Umbelliferae
70. Bupleurum 2500–4500
71. Heracleum 3000–4900
72. Pimpinella 3600–4200
73. Pleurospermum 3000–4500
74. Trachydium 3300–5500

Caprifoliaceae
75. Lonicera 3000–5100

Valerianaceae
76. Valeriana 3000–4200

Dipsacaceae
77. Morina 3000–4200
78. Scabiosa 2500–4200
79. Triplostegia 2500–4200

Compositae
80. Allardia 3900–5500
81. Anaphalis 1000–5000
82. Arctium 2500–3900
83. Artemisia 1000–5500
84. Aster 1500–5500
85. Brachyaster 3000–3900
86. Carduus 1800–3000
87. Centaurea 1500–3600
88. Chrysanthemum 3900–5100
89. Cnicus 2800–4200
90. Cousinia 3000–4200
91. Cremanthodium 3000–4900
92. Crepis 1300–3900
93. Doronicum 3000–4200
94. Echinops 1200–3900
95. Erigeron 2500–5500
96. Gnaphalium up to 3600
97. Gerbera 2600–3900
98. Hieracium 1800–3300
99. Inula 2800–4200
100. Jurinea 3300–4200
101. Lactuca 1200–5000
102. Leontopodium 3000–5500
103. Saussurea 3000–5500
104. Scorzonera 3000–4500
105. Senecio 4500–5100
106. Tanacetum 3000–5500
107. Taraxacum 3000–5500
108. Werneria 2500–5000

Ericaceae
109. Cassiope 3000–3900
110. Gaultheria 3000–3900
111. Rhododendron 2500–4900

Diapensiaceae
112. Diapensia 3000–4500

Plumbaginaceae
113. Acantholimon 3300–4500

Primulaceae
114. Androsace 3000–4900
115. Lysimachia up to 3000
116. Primula 1500–5100

Gentianaceae
117. Exacum up to 3000
118. Gentiana 2800–5100
119. Jaeshkea 2500–5100
120. Pleurogyne 3000–5500
121. Swertia 1500–3600

Boraginaceae
122. Arnebia 3000–3900
123. Cynoglossum 3000–3600
124. Echinospermum 2800–3600
125. Eritrichium 2500–4900
126. Lindelofia 3000–4500
127. Microula 4500–5100
128. Macrotomia 3000–4200
129. Myosotis 3000–5100
130. Onosma 3600–4200
131. Paracaryum 3000–3600
132. Rochelia 3600–3900
133. Solenanthus 3000–3900

Scrophulariaceae
134. Lagotis 3300–5500
135. Pedicularis 3000–5000
136. Picrorhiza 3000–4500
137. Scrophularia 1500–4200
138. Veronica 2500–5100

Labiatae
139. Calamintha 1800–4200
140. Dracocephalum 3000–5000
141. Elsholtzia 1500–5000
142. Hyssopus 2500–3600
143. Marrubium 2500–4900
144. Nepeta 2500–5100
145. Origanum 2500–3900
146. Perovskia 2500–3900
147. Phlomis 1500–3900
148. Scutellaria 2500–5000
149. Thymus 2500–4500

Chenopodiaceae
150. Atriplex up to 4200
151. Axyris 2500–4200
152. Chenolea 3300–4200
153. Chenopodium up to 4200
154. Corispermum 3000–4000
155. Eurotia 2500–4500
156. Haloxylon up to 4200
157. Kochia 2500–4200
158. Microgynoecium 3300–4500
159. Salsola 3300–4500

Polygonaceae
160. Oxyria 3600–5500
161. Polygonum 2500–5500
162. Rheum 3000–4500
163. Rumex 3600–3900

Elaeagnaceae
164. Hippophaë 2500–3600

Thymeleaceae
165. Stellera 3000–3300

Euphorbiaceae
166. Euphorbia 3000–5000

Haemodoraceae
167. Aletris 3000–4500

Irideae
168. Iris 2500–3600

Liliaceae
169. Allium 3000–5500
170. Eremurus 2500–4000
171. Fritillaria 3000–3600
172. Gagea 1200–3900
173. Lloydia 3600–5100
174. Polygonatum 3000–3900
175. Theropogon 1500–3300
176. Trillium 2500–3600

Juncaceae
177. Juncus 3000–5100

Aroideae
178. Arisaema 2500–4200

Cyperaceae
179. Carex 3000–5100
180. Cobresia 3000–5100
181. Scirpus 3000–4200

Gramineae
182. Bromus 3000–4500
183. Calamagrostis 3000–4500
184. Danthonia 3000–4500
185. Deschampsia 3000–5000
186. Deyeuxia 3000–5000
187. Elymus 3000–5500
188. Festuca 3000–5500
189. Glyceria 3000–5000
190. Poa 3000–5100
191. Stipa 3000–5100

It is seen from the foregoing table that the altitude range of most high altitude plants from the Northwest Himalaya lies between 3600 and 5500 m, though in many cases it is somewhat less and lies between 3900 and 4200 m. In some cases it extends beyond 5500 m and reaches almost up to 6000 m and descends to nearly 3000 m. Xerophytes and semi-xerophytes seem on the whole to have a decidedly wider range of altitude than most mesophytes and hygrophiles. In the Eastern and Central Himalaya, the xerophytes are, however, severely restricted to a much narrower belt of altitude than in the Northwest Himalaya. As may be expected, cryophytes have a relatively narrow altitude range. Their lower and upper limits of altitude are conspicuously higher than the general average. The width of altitudinal range thus

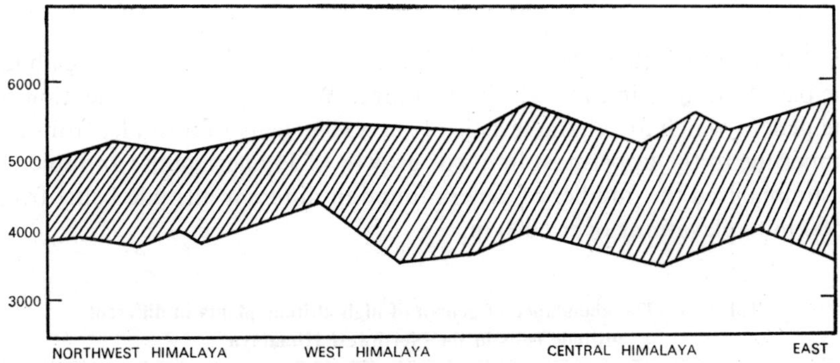

Fig. 41. East-west fluctuations in the mean widths of altitudinal range of distribution of high altitude plants in the Himalaya.

varies between 300 and 3000 m, but the general average is about 2000 m. The difference in the width of the altitude range is largely specific, but also dependent on other complex ecological characters of the plants. In certain families like Compositae, the seeds of which are dispersed widely by wind, the width of the altitudinal range is somewhat large and amounts to almost

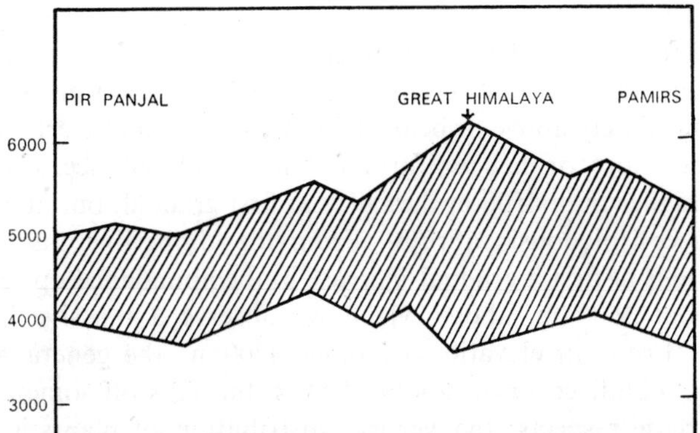

Fig. 42. The south-north fluctuations in the mean widths of altitudinal ranges of distribution of high altitude plants in the Northwest Himalaya.

2500 m. The width is also otherwise influenced by local topographical diversities, narrowing in gorges and transverse valleys and expanding on ridges. The mean width varies greatly from the east to the west (Fig. 41). The mean altitudinal range is wider in the eastern end than in the western end of the Himalaya. There is also a recognizable difference in the mean altitudinal width of plants as we proceed northward from the Pir Panjal Range over the Great Himalaya to the Pamirs (Fig. 42).

4. GENERA CHARACTERISTIC OF DIFFERENT ALTITUDE BELTS

The mean width of the altitudinal range does not tell us anything about the pattern of distribution within the range. We may now pay attention to the genera which are characteristic of successive belts of altitude, from the timberline to the uppermost limits of growth of Phanerogams. In table IV is summarized the number of genera found in different altitude belts, from 2500 m to elevations over 6000 m.

Table IV. The abundance of genera of high altitude plants in different altitude belts in the Northwest Himalaya

Altitude belt	Number of genera
2500–3000	90
3000–3600	170
3600–4200	160
4200–4500	110
4500–5000	90
5000–5500	40
5500–6000	10
Over 6000	2

We find from this table that between 80 and 90% of the genera of high altitude plants occur in the zone between 3600 and 4500 m. The elevated regions immediately above or below these limits are characterised by a progressive and more or less uniform fall in the abundance of genera. At elevations higher than 4500 m the fall is at first gradual, but above 5500 m it becomes conspicuously abrupt. From the timberline, as we ascend higher, there is at first a gradual increase in the abundance up to an elevation of about 3600 m, above which we observe a massive increase in abundance. From an elevation of about 4200 m the genera abundance seems to be stabilized up to about 4500 m, but falls off somewhat higher above. In these respects, the vertical distribution of plants is strikingly different from that of insects [75, 76]. In insects there is an abrupt and massive fall in abundance and variety of genera and species at the timberline and this fall continues undiminished higher up to the snowline. In plants on the other hand, there is a gradual increase in abundance at the timberline, then a massive increase, which is finally followed by a gradual fall near the snowline. For insects the timberline seems to be the threshold, and for plants it is the snowline that serves in this manner. This can have only one meaning viz., the plant communities are more stable than the insect communities at high altitude.

We may now briefly mention the plants which are characteristic of different altitudinal belts.

The genera characteristic of the zone immediately above the timberline (about 3000–3600 m) include *Anemone, Caltha, Delphinium, Isopyrum, Oxygraphis, Ranunculus, Trollius, Meconopsis, Stylophorum, Hypecoum, Alyssum, Braya, Capsella, Cheiranthus, Chorispora, Christolea, Cochlearia, Draba, Iberidella, Arenaria, Cerastium, Thylacospermum, Biebersteinia, Caragana, Guldenstaedtia, Oxytropis, Thermopsis, Parnassia, Trachydium, Allardia, Brachyaster, Cousinia, Cremanthodium, Doronicum, Jurinea, Leontopodium, Scorzonera, Tanacetum, Taraxacum, Cassiope, Diapensia, Acantholimon, Primula, Androsace, Arnebia, Microula, Macrotomia, Myosotis, Onosma, Rochelia, Lagotis, Pedicularis, Picrorhiza, Dracocephalum, Microgynoecium, Oxyria, Rheum, Rumex, Aletris, Fritillaria, Lloydia, Polygonum,* etc.

In the zone between 3600 and 4200 m the following genera may be considered as characteristic: *Aconitum, Adonis, Aquilegia, Callianthemum, Cimifuga, Stylophorum, Hypecoum, Alyssum, Barbarea, Capsella, Cardamine, Braya, Crambe, Draba, Parrya, Sisymbrium, Thlaspi, Gypsophila, Sagina, Silene, Erodium, Geranium, Guldenstaedtia, Sempervivum, Bupleurum, Pimpinella, Pleurospermum, Lonicera, Valeriana, Morina, Scabiosa, Triplostegia, Arctium, Brachyaster, Cnicus, Cousinia, Cremanthodium, Doronicum, Gerbera, Inula, Scorzonera, Werneria, Acantholimon, Androsace, Primula, Arnebia, Cynoglossum, Echinospermum, Macrotomia, Lindelofia, Onosma, Rochelia, Solenanthus, Picrorhiza, Hyssopus, Origanum, Perovskia, Phlomis, Axyris, Chenolea, Corispermum, Microgynoecium, Kochia, Rheum, Rumex, Stellera, Aletris, Eremurus, Juncus, Fritillaria, Polygonatum, Trillium, Bromus, Danthonia, Calamagrostis, Stipa,* etc.

The genera dominant at elevations up to 5500 m include the following: *Anemone, Caltha, Delphinium, Isopyrum, Oxygraphis, Ranunculus, Trollius, Meconopsis, Alyssum, Braya, Capsella, Chorispora, Christolea, Draba, Iberidella, Thlaspi, Arenaria, Cerastium, Lychnis, Thylacospermum, Biebersteinia, Caragana, Astragalus, Parochaetus, Thermopsis, Potentilla, Saxifraga, Sedum, Trachydium, Allardia, Erigeron, Leontopodium, Tanacetum, Taraxacum, Diapensia, Primula, Jaeshkea, Pleurogyne, Eritrichium, Microula, Myosotis, Lagotis, Pedicularis, Dracocephalum, Marrubium, Scutellaria, Oxyria, Polygonum, Lloydia, Allium, Carex, Cobresia, Deschampsia, Deyeuxia, Elymus, Festuca, Glyceria, Poa,* etc.

The genera characteristic of elevations above 5500 m are exemplified by *Anemone, Caltha, Delphinium, Isopyrum, Oxygraphis, Meconopsis, Alyssum, Braya, Chorispora, Christolea, Cochlearia, Draba, Iberidella, Parrya, Thlaspi, Arenaria, Cerastium, Lychnis, Thylacospermum, Biebersteinia, Astragalus, Caragana, Oxytropis, Thermopsis, Potentilla, Saxifraga, Sedum, Trachydium, Allardia, Cremanthodium, Artemisia, Erigeron, Leontopodium, Tanacetum, Taraxacum, Diapensia, Primula, Jaeshkea, Pleurogyne, Eritrichium, Microula, Myosotis, Lagotis, Pedicularis, Dracocephalum, Oxyria, Allium, Lloydia, Juncus, Carex, Cobresia, Deschampsia, Deyeuxia, Elymus,*

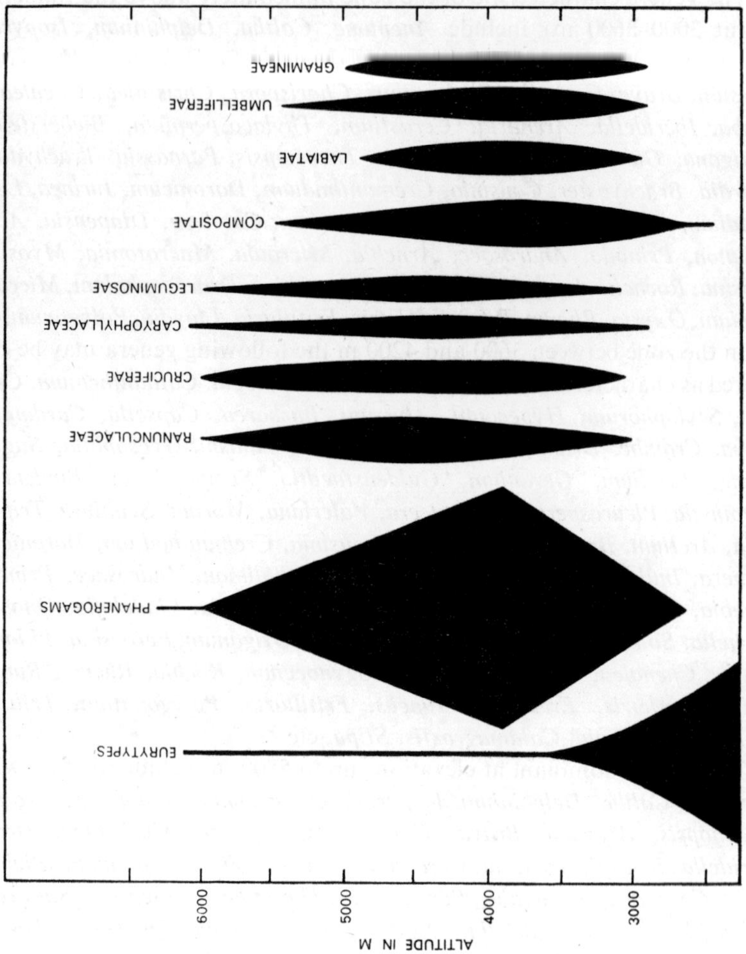

Fig. 43. Altitudinal fluctuations of abundance and dominance of the major high altitude plants in the Northwest Himalaya. Widely distributed eurytype plants, which often ascend into the high altitude ecosystem, show an abrupt fall in abundance from the timberline. The true high altitude Phanerogams, on the other hand, rapidly increase in abundance to attain a dominant position at altitudes of about 4000 m and become sparse only above 5000 m. Unlike in the case of insects, there is no massive fall in the abundance of high altitude plants at timberline, but on the other hand a progressive increase up to nearly 4000 m.

Festuca, *Glyceria*, *Poa*, etc.

Mention may also be made here of some of the plants, which are characteristic of the upper reaches of the forest zone, but which ascend often to extreme high elevations in the Northwest Himalaya. The typical examples belong to *Ranunculus*, *Thalictrum*, *Corydalis*, *Erysimum*, *Eutrema*, *Viola*, *Stellaria*, *Melilotus*, *Ononis*, *Parochaetus*, *Trifolium*, *Fragaria*, *Geum*, *Spiraea*, *Epilobium*, *Anaphalis*, *Artemisia*, *Aster*, *Carduus*, *Centaurea*, *Crepis*, *Echinops*, *Gnaphalium*, *Hieracium*, *Lactuca*, *Lysimachia*, *Exacum*, *Gentiana*, *Swertia*, *Scrophularia*, *Calamintha*, *Elsholtzia*, *Nepeta*, *Thymus*, *Atriplex*, *Chenopodium*, *Salsola*, *Euphorbia*, *Gagea*, *Theropogon*, *Scirpus*, etc.

5. MAJOR GROUPS

On the basis of the width of altitudinal range and the dominance at different altitude zones, three major groups of plants may be recognized viz., 1) plants which are characteristic of a transitional belt of altitude immediately above the timberline and generally not ascending to elevations above 3600–3900 m; 2) plants which are characteristic of a relatively wide belt of elevations between 3600 m and 5000–5500 m; and 3) plants which are characteristic of extreme elevated areas above 5500 m and reaching up to maximum limits of plant life.

This grouping is readily evident in the distributional pattern of certain families like Cruciferae, Leguminosae, Compositae, etc. (Fig. 43). As may be understood, the second group constitutes the largest elements which must also be considered as typically autochthonous to the high altitude ecosystem. The maximum number of plants, representing about 80–90% of those known at present, seem to be stabilized ecologically at altitudes lying between 3600 and 5500 m. Most climax communities and formations are also generally concentrated in the zone of 3900–4500 m. The plant communities of this zone are characterized by few pioneer formations (Fig. 44), but the insects in this zone, though rich in stable communities, contain many more pioneer communities than plants [75].

6. THE FRINGE ZONE PLANTS

Having discussed the width of altitude range and the genera characteristic of different altitude belts, we may now draw attention to some of the interesting plants found in the fringe zone. The ecologically transitional areas close to the extreme upper and lower limits of the altitude range constitute the fringe zone. Certain peculiarities of the fringe zone plants must be stressed here.

The fact that a plant is found growing in the fringe zone, at altitudes near the upper limit or even below the lower limits, does not necessarily mean this is its permanent establishment. This is particularly true of the plants growing at extreme high elevations of 6000 m and higher. To some extent it is also applicable to the high altitude plants occurring at elevations below 3000 m.

In nearly all cases, the fringe zone plants are readily recognized by the usually small number of individuals. The plants grow in small and greatly isolated pockets, as stray individuals. Many of these plants may sometimes be found blooming, but fruits and seeds are as a rule rarely if ever set. The colonies of the fringe are, strictly speaking, outside of their ecological optima and depend for their continued existence on the dispersal of seeds from the main autochthonous altitudinal belt. In the case of colonies about elevations of 6000 m, the source of seed year after year is from below, in fact from the zone between 3600 and 5500 m. In the case of pockets of high altitude plants found below the timberline, the source of seed is similarly from above, that is from the same belt between 3600 and 5500 m. It is from this centre of their main altitudinal autochthonous zone that continuous upward or downward dispersal of seeds tends to very gradually widen the

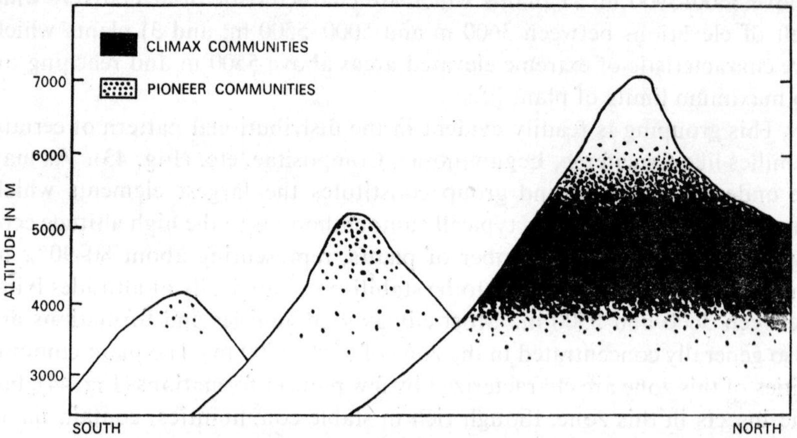

Fig. 44. Altitudinal frequencies of pioneer and climax communities of high altitude plants in the Northwest Himalaya. The Himalayan System in the region is shown in simplified diagrammatic sectional view. The concentration of maximum of climax and ecologically stable communities of the high altitude plants in the zone between 3600 m and 5500 m is indicated by the dark and denser shaded area. This is also the zone in which the largest numbers of high altitude genera and species occur. Both above an elevation of 5500 m and also below 3600 m, the marked fall in numbers and the sparseness of climax communities are indicated by lighter and open shading; these fringe zones are areas of increasing pioneer communities. At elevations above 5500 and reaching above 6000 there is a dominance of pioneer communities of plants, gaining in altitude from the zone below. At elevations below 3600 m the dominance of pioneer communities is partly from colonies from above, descending, and partly from colonies ascending from the timberline belt.

width of the altitudinal range. Favourable pockets outside of the permanent altitude belt are generally the margins of melt water streams, screes along the upper valleys etc.

While the typical high altitude plants are autochthonous within the belt between 3600 and 5500 m, in which they regularly flower and set seeds, the fringe zone, about a hundred metres wide, both above 6000 m and below 3000 m, is not permanent. In the autochthonous zone the plant communities are permanent, stable and ecologically approach climax conditions (Fig. 44). Outside of the fringe zone, the colonies of plants are either pioneers, which are actively prograding either upwards or downwards as the case may be, or they are merely accidental. Maximum ecological stability of the high altitude plants is observed in a relatively narrow belt, about 300–900 m wide, lying between 3600 and 4500 m. Both above and below these altitudes, there is increasing ecological instability.

It is not uncommon to come across stray individuals of typical valley plants or even of eurytype lowland plants growing at high altitudes, often as high as 4500 m, particularly on the southern slopes of the mountains. This must not be interpreted as evidence of a wide altitude range of these species. The presence of such plants in the high areas is to be traced to transport of nutlets, seeds, etc., by the updraft air currents, during summer from the valleys or even from the distant plains [75]. They differ from the well known wind-blown derelicts, so commonly observed on snowfields on the Himalaya [75, 76, 77] in the fact that atmospheric cold of the high altitude does not prevent them from germinating, although the plants generally remain starved and stunted on account of the cold, wind and other unfavourable conditions. None of these plants bloom.

Although not among the component species of the high altitude ecosystem, some of these visitor species appear to be in the process of gradually being assimilated as regular members. In their general appearance and habits most of them are almost indistinguishable from the true component elements. The individuals of these plants are nearly all densely hairy, much branched, tufted and stunted, with relatively small or dwarfed leaves and short internodes. These characters are strong evidence of recent enrichment of the high altitude plant communities by additions from other neighbouring ecosystems. The conditions for such incorporations of members from outside appear to be particularly favourable in the Northwest Himalaya. The commonest examples of such recent incorporations are met with in Caryophyllaceae, Compositae and Chenopodiaceae.

7. SUMMARY

This chapter presents an outline of the striking peculiarities of the vertical distribution of high altitude plants in the Northwest Himalaya. In marked contrast to the distribution of insects, there is no abrupt and massive fall either in the abundance or variety of plants at the timberline. While no doubt trees and large shrubs and many herbaceous plants, which are typical of the forest floor, disappear at the timberline, their place is taken by very large numbers of newer types, which are not found at lower elevations. There is even a marked rise in the general abundance of genera and species immediately above the timberline. This increase continues almost up to elevations of 4500–5000 m, above which only we begin first to observe a gradual thinning off of numbers and variety.

The highest altitude at which different plants occur in different geographical divisions of the Himalaya and the mean width of altitudinal range of distribution are discussed in some detail. The highest altitude at which a Phanerogam has so far been observed in the Himalaya is 6300 m.

Hygrophiles and mesophytes ascend much higher in the Eastern Himalaya than in the Northwest, where the xerophytes and subarid-frigid steppes elements are not only dominant but also occur at much higher elevations.

The pattern of altitudinal distribution of plants is greatly complicated in the Northwest Himalaya by facts of horizontal distribution, so that the two cannot be separated. A tridimensional picture is needed to present a satisfactory account of the distribution of high altitude plants in the Northwest Himalaya.

On the basis of the altitudinal distribution, three major groups of high altitude plants are recognized: 1) the plants which are typical of the zone immediately above the timberline; 2) the plants typical of the zone extending from 3600–3900 m to 4500–5000 m and 3) plants typical of elevations 5500–6000 m and higher.

The greatest bulk of the high altitude plants occurs in the zone between 3600–3900 m and 4500–5000 m. Both above and below these limits, their abundance diminishes. The significance of this peculiarity is explained by the conclusion that the high altitude plants differentiated, evolved and diversified within this zone and may, therefore, be considered as truly autochthonous to it. It is also within this zone that the maximum numbers of ecological stable communities, formations and associations of not only plants, but also of high altitude insects, approaching climax status, are found. The few isolated plants found growing more or less frequently at extreme elevations, far above these limits, as well as those that appear near the timberline, are pioneers and members of prograding colonies from this zone. There is also evidence of recent incorporation of members from outside ecosystems into the high altitude ecosystem.

PLATES

Plate II. The Pir Panjal Range.

1. The Pir Panjal Range. The upper reaches of the forest zone, with *Cedrus deodara*, *Abies pindrow*, etc. in the foreground and the treeline in the distance.

2. The Pir Panjal Range, the mountain slope near the treeline, with *Alnus nitidus* near glacial stream beds.

Plate III. The timberline on the Himalaya.

1. The timberline zone on the southern slopes of the Northwest Himalaya, with *Betula utilis* and bushes of *Rhododendron campanulatum*.

2. View of an inner valley of the Northwest Himalaya, immediately above the timberline, with the Great Himalayan Range on the left and the Pir Panjal Range on the right.

Plate IV. The Himalaya rises as a mighty wall of rock and snow.

1. The Great Himalaya, rising as a mighty wall of ice, snow and barren rock, beyond the many parallel Lesser Himalayan Ranges. In the near foreground is the Siwalik Range, covered by tropical and subtropical forests. Beyond it, in the middle distance, are other mountain ranges, covered by temperate forests. Further to the north in the background are mountains with open vegetation, above which rises the main range, seemingly to reach the sky.

2. The timberline zone in the interior ranges of the Northwest Himalaya, markedly different from the southern slopes of the Pir Panjal Range, generally subarid and reminding one of the landscape in the high Pamirs.

Plate V. The arid regions of the Himalaya.

1. One of the inner and higher valleys in the Northwest Himalaya, in which no place is below an elevation of 3300 m.

2. The timberline zone in the arid extreme western end of the Northwest Himalaya, with stunted bushes of *Betula*, spruce and juniper.

Plate VI. The Great Himalaya.

1. The Great Himalaya, in the region of elevations above 5500 m, comprising the nival and aeolian zones.

2. The Great Himalaya, at an elevation of about 5000 m—a world of rock, glacier ice, crevasses and snow. the home of some incredibly delicate and many other remarkably hardy cryophyte plants.

Plate VII. Some petrophytes.

2. *Saxifraga ciliata* spreading its leaves from persistent rootstock during early July on rock surface in a moderately moist locality at 3300 m on the south slope of the Pir Panjal Range.

Plate VIII. Petrophytes, mesophytes and hygrophites.

1. A subarid locality, with petrophite formations of crusty and brightly coloured lichens on the exposed rock surface and spreading mats of delicately coloured flowers of the aromatic *Thymus serphyllum* in between the boulders, at an elevation of 4200 m during July; this is a part of the *Artemisia*-steppes.

2. Hygrophile locality, with thick moss cushions, containing an abundance of *Primula denticulata*, *Caltha* and other mesophytes in bloom on the edge of melt-water cascade on boulders, at an elevation of 3900 m on the northern slopes, during June.

Plate IX. Alpine meadows.

1. A high alpine zone grassy meadow, at an elevation of 4000 m, with *Heracleum*, *Pimpinella*, etc. in bloom during June, a typical subarid habitat.

2. A high alpine zone grassy meadow at an elevation of 4000 m, with Cruciferae, *Valeriana* and some Umbelliferae and *Erigeron*, a mesophile habitat.

Plate X. Alpine meadows.

1. High alpine grassy meadow, at an elevation of 3900 m, with masses of *Aster molliusculus* in bloom during June-July.

2. Moist alpine grassy meadow at elevations between 3600 and 3900 m, with *Aster*, *Polygonum*, *Geranium*, *Potentilla*, etc. in bloom during May-June.

Plate XI. Some plants in bloom.

1. A mat of the delicate and brightly coloured flowers of *Androsace microphylla*.

2. A subarid and exposed high alpine meadow at an elevation of 4000 m, with flowers of *Erigeron*, attracting flies and beetles, during June.

Plate XII. Alpine meadows.

1. A moist high alpine zone meadow, near a snow edge and melt-water stream, with an abundance of *Anemone rivularis* and some *Caltha palustris*.

2. *Gypsophila* in flowers at an elevation of 3900 m.

Plate XIII.

1. *Androsace* in a sheltered rock crevice blooming at an elevation of 3600 m during May.

2. *Gerbera kunzeana*.

Plate XIV. Some typical plants.

1. *Carduus*, 2. *Saussurea gossypiphora*, 3. *Rhododendron campanulatum*, 4. *Saussurea obovallata*, 5. Arid slopes at an elevation of 4500 m covered by cushions of dwarf plants, often of considerable age, 6. *Barbarea.*

Plate XV. Meadows and moss cushions.

1. A subarid high meadow, with dwarfed *Lactuca*.

2. Sheltered moist cushions of moss, growing in great luxuriance, at an elevation of 4200 m during June.

Plate XVI.

1. *Taraxacum* at an elevation of 3500 m.

2. *Chorispora sabulosa* at an elevation of 3500 m.

Plate XVII. Photographs of herbarium specimens.
1. *Thermopsis barbata.*

2. *Fritillaria roylei.*

Plate XVIII.

1. *Allium jovanianum* with its beautiful stellate delicate white flowers at an elevation of 4200 m on alpine meadows during June.

2. *Androsace sempervivoides*, with its typical rosette of leaves and delicate flowers during June at an elevation of 4200 m on the northern slopes of the Pir Panjal Range.

Plate XIX. Plants and insects at high altitude.
1. *Sedum quadrifidum.*

2. Hairy muscoid flies visiting the flowers of *Primula* during the hours of bright sunshine.

3. Syrphid flies visiting the flowers of *Primula*.

Plate XX. Alpine meadows.

1. A high alpine grassy meadow, at an elevation of 3900 m, with *Polygonum*, *Potentilla* and other plants in bloom.

2. Masses of flowers of *Androsace sempervivoides* attracting hairy-bodied flies at an elevation of 4200 m during May-June on the Great Himalaya.

Plate XXI. Alpine meadows.

1. A high alpine grassy meadow at an elevation of 3600 m, with a massed carpet of brightly coloured flowers of *Thermopsis barbata*, *Fragaria*, etc. on the Great Himalaya, during late June.

2. An alpine meadow on the south slope of the Pir Panjal Range, at an elevation of 3600 m, a carpet of the flowers of *Caltha*, *Trollius*, etc. during summer.

1

2

Plate XXII.

1. *Primula floribunda*, with its delicate yellow flowers at the tip of a long scape, growing in a relatively moist locality, at an elevation of 3600 m in July.

2. *Dracocephalum diversifolium* in bloom during May at an elevation of 3300 m on the southern slopes of the Pir Panjal Range.

Plate XXIII.

1. *Caltha* in full bloom in moist localities during May.

2. *Anemone* flower at the edge of a melt-water stream at 3900 m on the northern slopes of the Pir Panjal Range.

Plate XXIV.

1. An alpine meadow at an elevation of 3900 m, with *Sedum quadrifidum*, with bright red flowers in the centre, and some *Caltha*, *Potentilla*, etc. in the background.

2. A flower of *Cnicus* visited by the bumble bee *Bombus* during the hours of bright sunshine in late June, in a semi-arid locality.

Chapter VI

Phytogeography

1. INTRODUCTION

THE peculiarities of the distribution of mountain plants are fairly well known and have been described in detail by several authors from different parts of the world. The distribution of the Himalayan plants, both from the floristic and vegetational points of view, constitutes an interesting chapter of phytogeography.

Though a number of botanists like Clarke [20], Hooker [59] and in more recent years Gupta [49, 50, 51], Hara [52], Kihara [64], Landon [71], Marquand [79], Meusel [85], Meusel & Schubert [86], Meher-Homji [80, 81], Schweinfurth [104], Troll [116] and others mentioned in the Bibliography have made valuable contributions to the phytogeography of the Himalaya, the emphasis has so far been largely on the forested slopes of the Himalaya and also in particular to the Nepal region. Most botanists have no doubt recognized the Northwest Himalaya as phytogeographically distinct, in parts at least, from the rest of the Himalaya. Very few of them have, however, devoted any attention to the phytogeography of the western end of the Himalaya. There is also much confusion regarding the precise boundaries of the major phytogeographical divisions, proposed by various authors, and of the affinities of the Aimalaya. Indeed our knowledge of the floristic and vegetational phytogeography of the Himalaya as a whole is extremely fragmentary. We also know practically nothing about the phytogeography of the elevated regions, above the timberline, in the Northwest Himalaya.

An attempt is made in this chapter to discuss, in brief outline, the floristic composition and geographical affinities of the plants of the high altitude ecosystem of the Northwest Himalaya. It must be pointed out, however, that in our present state of knowledge of the flora of this region, this discussion is necessarily bound to be rather sketchy. It seems possible only to outline certain broad facts, based on the general ecological characters and peculiarities of vertical distribution, described in foregoing chapters. These facts form the basis for certain tentative, though fundamentally important, conclusions on the sources and evolution of the high altitude plants of the region. It is hoped that extended explorations in future should explain many obscure facts and substantiate the broad ideas presented here.

2. GEOGRAPHY OF THE HIGH ALTITUDE PLANTS

It seems best to prefix the phytogeographical analysis of the high altitude ecosystem of the Northwest Himalaya by a brief review of the known geographical distribution, outside of the region, of the genera and species of high altitude plants. The genera are then grouped on the basis of the pattern of their distribution. The distributional data, summarized here, are well known in literature and are generally widely accepted.

RANUNCULACEAE

The genus *Anemone* is widely distributed in the north temperate and arctic regions of Europe, Asia and North America and several species are known from the Alps, Siberia, the Altai Mountains, the Central Tien Shan, the Alai-Pamirs, the Himalaya and the Western Ghats. Of the high altitude species from the Northwest Himalaya, *A. albana* extends from Armenia to North Asia and the Altai Mountains. *A. rupicola*, *A. rupestris* and *A. obtusiloba* extend the whole length of the Himalaya. *Aquilegia* is known from the temperate Europe, Caucasus, Asia and North America. *Actaea* is known from the Arctic north of Europe, Asia and North America. *Aconitum* is known from the north temperate areas of Europe, Asia and North America and extends also into circumpolar region. *Adonis* is known from Europe and Asia. *Caltha* is distributed in temperate Europe, Asia and North America. *Cimifuga* extends from eastern Europe to Siberia. *Callianthemum* is distributed in Europe, the Hindu Kush, the Himalaya and Siberia. *Delphinium* is known from Europe, North Africa, the whole of the Himalaya and the Kashi Hills in Assam. *Isopyrum* extends from eastern Europe to Siberia. *Ranunculus* is widely distributed in the temperate and arctic areas of Europe and Asia and extends to North America and the Western Ghats. Among the high altitude species of this genus from the Northwest Himalaya, *R. cymbalariae* is known also from Iran, Siberia and North America; *R. pulchellus* is known from Afghanistan, Siberia and Mongolia; *R. affinis* from northern Asia and North America and *R. nivalis* is distributed in Europe, northern Asia and North America. The genus *Oxygraphis* is known also from Siberia. *Thalictrum* is rather widely distributed in the temperate regions in Europe, Asia, North America and is also known from the high mountains of tropical areas, including Java and the Western Ghats. *Trollius* is distributed in temperate and arctic Europe and Asia.

Papaveraceae

The genus *Stylophorum* is known from Europe and temperate Asia and *Meconopsis* occurs on the mountains of the north temperate lands.

Fumariaceae

The genus *Hypecoum* extends from western Asia to the Central Tien Shan and the whole of the Himalaya and *Corydalis* has a wide distribution in Europe, Asia and North America and is particularly abundant in temperate areas.

Cruciferae

The genus *Alyssum* is distributed in southern Europe and extends through western Asia to Middle Asia, Central Asia and Siberia. *Arabis* is known from Europe, North and East Africa, western Asia through Middle and Central Asia to North Asia and North America. *Barbarea* is known from Europe, western Asia, North, East and South Africa and Middle Asia. *Braya* is distributed in temperate and arctic Europe, Asia, the Alps, Middle Asia, Siberia, the Altai and the Andes. The high altitude species *B. rosea* from the Northwest Himalaya is known also from the Altai Mountains and Siberia. *Capsella* is widely distributed in the north and south temperate lands and occurs in Europe, North Africa and Middle Asia. The common *Capsella bursa-pastoris* is widely distributed in North Africa, Europe and extends to Middle Asia. *Cardamine* is widely distributed in the north and temperate areas. *Cheiranthus* extends from North Africa to Europe, West Asia eastwards to arctic northern America. *Chorispora* extends from West Asia to Middle Asia and occurs also in the whole of the Himalaya. *Christolea* is known from the Himalaya and western Tibet. *Cochlearia* is a genus of the north temperate. *Crambe* occurs in Europe, Caucasus, Iran, western Tibet and the whole of the Himalaya. The genus *Draba* has an interesting circumpolar distribution and also occurs in temperate Europe, Caucasus, Iran, Middle Asia, the Himalaya and the Andes. The high altitude species *D. alpina* from the Northwest Himalaya is known also from the Taurus Mountains, northern Europe, Middle Asia and the Rocky Mountains in North America and *D. incana* has a circumpolar distribution. The genus *Erysimum* is known from Europe, North Africa, Caucasus and Siberia. *Eutrema* is an arctic genus and occurs in Siberia and the Himalaya. *Iberidella* occurs, in addition to the Himalaya, in western Asia also. *Parrya* occurs in temperate, arctic and circumpolar areas of Europe, Asia and North America; is known from Afghanistan, the Altai Mountains, the whole of the Himalaya and the Alai-Pamirs. *Sisymbrium* is widely distributed in temperate Europe and Asia, Central European mountains, North Africa, arctic North America, Siberia, the Altai Mountains, Hindu Kush, the Alai-Pamirs and the entire Himalaya. The distribution of *Thlaspi* extends over the whole of temperate Europe and Asia, the Alps, North America and the Andes.

VIOLACEAE

The genus *Viola* has a wide distribution in the temperate regions of the world and is known from Europe, western and northern Asia, Middle Asiatic mountains, the Himalaya, the hills of Assam (Khasi and Jaintia Hills), the Satpura, the Eastern and Western Ghats and the mountains of Sri Lanka.

CARYOPHYLLACEAE

Silene is distributed in northern Asia, the mountains of Middle Asia, the Himalaya and the Western Ghats. *Lychnis* occurs in temperate and arctic regions and has a circumpolar distribution, with a number of species in Europe, Asia and North America. Several species occur in Siberia, the Himalaya and western Tibet. The genus *Cerastium* is likewise common in the arctic and temperate areas and is recorded from the Alps, Caucasus, Hindu Kush, Pamirs, Tien Shan and the Himalaya; it is also known from the Western Ghats. *Danthonia barbatus*, *D. chinensis* and *D. caryophyllus* and *Saponaria*, which are not strictly high altitude plants in the Northwest Himalaya but occur sometimes above the timberline, are known also from the Central Tien Shan. Species of *Stellaria* are known to be widely distributed in Europe, Asia, Greenland, Siberia, Middle Asiatic mountains, the Himalaya and the Western Ghats and Sri Lanka hills. The genus *Arenaria* extends from the Caucasus through the Hindu Kush-Karakoram complex to the Himalaya, Pamirs and the Central Tien Shan and is also known from the Western Ghats. *Sagina* is common in north and south temperate areas, in the Himalaya and the Western Ghats. Species of *Thylacospermum* appear to be restricted to the Middle Asiatic mountains and Songoria. The genus *Gypsophila* seems to be largely confined to the Himalaya, Tibet and Middle Asia.

TAMARISCACEAE

Myricaria occurs in Europe, the mountains of Middle Asia and drier parts of the Himalaya.

GERANIACEAE

The genus *Biebersteinia* extends from Siberia through the Altai Mountains to the mountains of Middle Asia. *Oxalis* is widely distributed throughout Europe, North Africa, northern Asia and North America, Middle Asia, the Himalaya and the plains and hills of India. *Geranium* is also widely distributed in Europe, western Asia to northern Asia, Siberia the Altai Mountains, Central Tien Shan, the Pamirs, Hindu Kush, the Himalaya and the Western Ghats. *Geranium pratense*, found in the high altitude ecosystem of the Northwest Himalaya, is known also from Europe, northern Asia and the Middle Asiatic mountains. The genus *Erodium* is known from Europe, North Africa, Siberia, Middle Asiatic mountains and the Himalaya.

LEGUMINOSAE

The genus *Astragalus* is widely distributed in Europe, northern Asia, Siberia, Middle Asia and western Asia and North Africa. Species of *Caragana* extend from the Altai Mountains to the Hindu Kush, Dahuria and Tibet. *Caragana pygmaea* of the high altitude ecosystem of the Northwest Himalaya is also known from Afghanistan. The small genus *Guldenstaedia* seems to be confined to the Himalaya and Middle Asiatic Mountains. *Melilotus* is widely distributed in Europe, Siberia, Middle Asiatic Mountains, the Himalaya and the Western Ghats. *Melilotus alba*, found in the high altitude ecosystem of the Northwest Himalaya, occurs as a winter weed in the plains of northwest India. *Ononis* occurs in Europe, widely in the Mediterranean steppes, in Siberia and Tibet. *Oxytropis* occurs on the mountains of Europe, in Siberia and on the mountains of Middle Asia. *Oxytropis lapponica* of the high altitude ecosystem of the Northwest Himalaya is known also from European mountains, from Siberia and Middle Asiatic mountains. *O. immersa* occurs also in the Pamirs. *Parochaetus* occurs on the Himalaya and on tropical mountains of Burma, Java, Nilgiri Hills and Sri Lanka hills. *Thermopsis* is known from Middle Asia, China, Japan and North America. *Trifolium* is widely distributed in Europe, Asia and North America and is also known from the Western Ghats and the mountains of Tropical Africa.

ROSACEAE

Alchemilla vulgaris is known not only from the Northwest Himalaya, but also from Iran, Middle Asia, northern Asia, and northern Europe and arctic North America and Greenland; it has circumpolar and boreal-alpine distribution. The genus is represented as a Pleistocene relict on the high

Western Ghats in South India. *Fragaria* is widely distributed on the mountains of the north temperate areas and occurs also in South America. *Geum* is known from Europe, Siberia and the Middle Asiatic mountains; *Geum urbanum* of the Northwest Himalaya is also known from Europe and Siberia. *Potentilla* is one of the widely distributed high altitude plants and occurs in Europe, Asia, North America and has a circumpolar range. Several of the high altitude species from the Northwest Himalaya are also known to occur on other boreal mountains. *Potentilla sibbaldi*, *P. fruticosa* and *P. bifurca* occur, for example, in Europe, northern Asia, the Altai and Middle Asiatic mountains. *P. solessovii* occurs in Altai and the Central Tien Shan. *P. anserina* is known from north Asia, North America, Europe and Iran. *P. bifurca* occurs also in Mongolia, Altai, Caucasus, Taurus, Songoria, China and North America. *P. serica* occurs in Afghanistan, Armenia, Pamirs, Songoria, northern China and boreal North America. *P. gelida* occurs in Dahuria, Songoria, the Central Tien Shan and the Pamirs. *P. nivea* extends north to the Central Tien Shan, Pamirs and westwards to the Caucasus and the Alps and is indeed boreal-arctic in Europe, Asia and North America. The genus *Spiraea* is known from the Altai Mountains, north China, Japan, Tien Shan, Pamirs, the Himalaya, western Europe and western North America.

SAXIFRAGACEAE

Parnassia is known from Japan, the Altai Mountains and through the mountains of Middle Asia to the Himalaya and the Western Ghats and the Khasi Hills. The genus *Saxifraga* is arctic-alpine in its distribution and several species are from Siberia, North America, Europe and the Middle Asiatic mountains, the Himalaya, Satpura and the Western Ghats. *Saxifraga sibirica* from the Northwest Himalaya is known also from the Altai Mountains, the Central Tien Shan and the Caucasus. *S. cernua* extends to the Pamirs and the Central Tien Shan and the Alps and is common in the north temperate and arctic areas of Europe, Asia and North America. *S. hirculus* is known from the Caucasus, the Pamirs, Central Tien Shan and the Alps.

CRASSULACEAE

The species of *Sedum* have wide distribution in arctic-alpine and circumpolar areas in Europe, Asia and North America and some species are known also from Mt. Kenya in Equatorial Africa and from Peru. *Sedum rhodiola* from the Northwest Himalaya occurs also in Europe, North Asia and North America, *S. quadrifidum* occurs in arctic Russia, Siberia and *S. ewersii* in Siberia, Songoria, the Pamirs and the Central Tien Shan. *Sem-*

pervivum is largely a Himalayan genus.

Onagraceae

Epilobium is widely distributed in Europe, arctic Asia and North America and occurs even in New Zealand and on the mountains of tropical lands. *E. angustifolium* from the Northwest Himalaya is known also from western Asia, Europe and North America. *E. latifolium* occurs in Dahuria, arctic Asia, Europe and America. *E. palustre* occurs in North Asia, Europe and America. *E. roseum* is known from western Asia and Europe.

Umbelliferae

Bupleurum extends from Europe eastwards to the arctic North America and occurs on the mountains of Middle Asia, the Himalaya, Western Ghats and the hills of Sri Lanka. *Heracleum* is known from Europe, North Africa, Middle Asia, the Himalaya and the Western Ghats. *Pimpinella* is known from Europe and temperate Asia and the Western Ghats and Khasi Hills. *Pleurospermum* occurs in the Himalaya and *Trachydium* extends to western Tibet.

Caprifoliaceae

Lonicera occurs in the north temperate areas and is also known from Siberia, Middle Asia and the Western Ghats. *Lonicera hispida* from the Northwest Himalaya is known from the Pamirs and Siberia. *L. microphylla* occurs on the Altai Mountains and Central Tien Shan and the Pamirs.

Valerianaceae

Valeriana is known from Europe, Caucasus, Middle Asia, the Himalaya, Khasi Hills, mountains of Java, Western Ghats and Sri Lanka hills and Mt. Kenya in East Africa. *Valeriana dioica* from the Northwest Himalaya is known also from Europe.

Dipsacaceae

Of the three genera occurring in the Northwest Himalaya *Scabiosa* is known also from Europe, Africa, western Asia, Arabia, Armenia, the Hindu Kush and Middle Asia. *Morina* seems to be restricted to western and Middle Asia. *Triplostegia* is distributed in the temperate and alpine zones of the Himalaya.

COMPOSITAE

Allardia is known from Middle Asia and western Tibet. *Anaphalis* is widely distributed in the plains and mountains of Asia and occurs also in North America; some of the species placed under *Helichrysum* from East African mountains may also belong to this genus. *Arctium* is distributed in temperate Europe, the Mediterranean region and Asia. *Artemisia* is widely distributed in Europe, North Africa, North, West and Middle Asia, Siberia, Altai Mountains, Mongolia, Caucasus, Tien Shan, Pamirs, Hindu Kush, the Himalaya, the Western Ghats and some species are known to extend to South America. Of the high altitude species from the Northwest Himalaya, *A. salssoloides* is known also from the Caucasus; *A. decurrens* occurs in Tibet, Afghanistan, West Asia and southern Russia and *A. maritima* occurs in coastal Europe and North Asia. The genus *Aster* has an interesting circumpolar distribution in Europe, Asia, North America and occurs besides in China, Siberia, the Altai Mountains, Tien Shan, Pamirs, Hindu Kush, western Tibet and the Himalaya. Of the high altitude species from the Northwest Himalaya, *A. heterochaeta* is known from the Altai Mountains and *A. alpinus* is widely distributed. The genus *Brachyaster* seems to be peculiar to the Himalaya and Middle Asia. *Carduus* is known from Europe, North Africa, Mt. Kenya, northern Asia, the Central Tien Shan and the Pamirs. *Centaurea* is widely distributed in Europe, western Asia and sparsely in North and South America, the Hindu Kush, the Himalaya, Caucasus and the Nilgiri Hills. *C. depressa* from the Northwest Himalaya extends westwards to Afghanistan and Caucasus. *Cnicus* is distributed in temperate lands and extends to north Asia and western Europe and occurs also on the south Indian hills. *Cousinia* is mostly Middle Asiatic but extends to western Asia also. The genus *Cremanthodium* is apparently confined to the Himalaya. *Crepis* occurs in Europe, Siberia, Mongolia, the Central Tien Shan, Pamirs, Himalaya, Khasi Hills and the Western Ghats. *Doronicum* is distributed in Europe and Middle Asia. *Echinops* has a Mediterranean distribution, particularly in the semi-arid areas and occurs in Europe, North Africa, tropical Africa, the Hindu Kush, Pamirs, Central Tien Shan, Japan, the semi-arid western parts of the Indo-Gangetic Plains of north India and the dry areas of south India (Deccan). *E. echinatus* occurs, for example, not only in the Northwest Himalaya, but also in the plains of the northwest India and extends westwards to Afghanistan. *Erigeron* is widely distributed in the north temperate areas and occurs in North America, Khasi Hills and the Western Ghats. *Erigeron alpinus* from the Northwest Himalaya is also known from the Central Tien Shan. *Gnaphalium* is known from Mt. Kilimanjaro, Mt. Kenya and the south Indian Hills. *Gerbera* occurs on the mountains of Africa, India Burma and China. *Hieracium* is distributed in western Europe and extends to north Asia, North

America, South Africa, Caucasus, Iran, Pamirs, the Himalaya and the Andes. *Hieracium umbellatum* from the Northwest Himalaya extends down to the lower elevations within the forest and westwards to Iran, Caucasus and Europe and northwards to the Pamirs and northern Asia and North America. *Inula* is distributed in Europe, Africa and Middle Asia. *I. rhizocephala* from the Northwest is known from the Pamirs also. *Jurinea* extends from Europe across western Asia and the Hindu Kush to Middle Asia. *Lactuca* is widely distributed in Europe, western Asia, Middle Asia, Siberia, the Pamirs, Central Tien Shan, the Himalaya, Hindu Kush, the Western Ghats, Khasi Hills and the mountains of Java and Sumatra. Of the high altitude species from the Northwest Himalaya, *L. orientalis* is known in western Tibet, Syria and Armenia; *L. scariola* occurs in Europe and Siberia and *L. tartarica* in Middle Asia and Siberia. *Leontopodium* is known from the Alps, Pamirs, Tien Shan and the Himalaya. The interesting genus *Saussurea* extends from the Altai Mountains through the Central Tien Shan and the Pamirs to Songoria, China, Japan, the Himalaya, Hindu Kush and the mountains of Burma. *S. sorocephala* occurs also in the Altai Mountains, Central Tien Shan and the Pamirs. *Scorzonera* is known to extend from Europe, North Africa and western Asia across Caucasus, Middle Asia to Mongolia. *S. divaricata* is known from Mongolia and western Tibet. The distribution of *Senecio* extends from Europe eastwards to the Himalaya, Pamirs, Tien Shan, southwards to Mt. Kilimanjaro, Mt. Kenya and the hills of south India, Khasi Hills and the mountains of Burma. *S. coronopifolium* from the Northwest Himalaya extends westwards through Afghanistan to Europe, as far as Spain. *Tanacetum* occurs in north temperate regions and is known from the Altai Mountains, the Central Tien Shan and Pamirs, Hindu Kush and the Himalaya. *Taraxacum* is distributed in cold and temperate areas. *Werneria* is recorded from North Africa and the Andes.

ERICACEAE

The genus *Cassiope* occurs in the cold north and in alpine zones of the Himalaya. *Gaultheria* is characteristic of tropical mountains and is known from the Andes, Himalaya, Western Ghats, etc. The distribution of *Rhododendron* covers the Alps, the Himalaya and the tropical mountains.

PLUMBAGINACEAE

Acantholimon is known from southern Europe, western Asia, Iran, the Pamirs, Central Tien Shan and the Northwest Himalaya.

DIAPENSIACEAE

Diapensia is known from Europe, Japan, the Himalaya, Middle Asia and North America.

PRIMULACEAE

The genus *Androsace* is widely distributed in Europe, the Alps, Caucasus, Hindu Kush, Himalaya and northern Asia. Some of the high altitude species from Northwest Himalaya like *A. septentrionalis* occur in North Asia, Caucasus and Europe; *A. sempervivoides* in Afghanistan and *A. villosa* in Caucasus, Alps, Asia Minor and Middle and North Asia. *Lysimachia* is widely distributed in Europe, China, Japan, Central Tien Shan, the Khasi Hills and the Western Ghats. The genus *Primula* is one of the well known circumpolar and arctic-alpine plants found in Europe, Asia and North America and some of the Northwest Himalayan species extend westward to the Hindu Kush and northwards to the Pamirs, Central Tien Shan, Altai Mountains, Siberia and North America. A few of the Himalayan species are known from the Khasi Hills also.

GENTIANACEAE

The genus *Exacum* is widely distributed in Malaya, China, India and Africa; *E. trigonum*, which is common nearly throughout the plains of India as a winter weed, ascends the Himalaya and occurs also on South Indian mountains. *Gentiana* occurs in Europe, Asia, North America and the Andes. Some of the species from the Northwest Himalaya extend to cold boreal areas. *G. borealis* occurs, for example, in Northwest America; *G. pecumbens* in Dahuria, Pamirs, Siberia and *G. dahurica* in Dahuria, Baluchistan, Afghanistan and Songoria. *Jaeshkea* seems to be restricted to the Himalaya. *Pleurogyne* is known from the Alps, Caucasus, Pamirs, Central Tien Shan, Hindu Kush, western Tibet, North America and the Western Ghats. The genus *Swertia* occurs on the mountains of Europe, the Himalaya, East Africa (Mt. Kenya), Middle Asiatic mountains, Khasi Hills, Deccan and the mountains of Sri Lanka.

BORAGINACEAE

The genus *Arnebia* is known from North Africa, western Asia, Middle Asia and the Himalaya. The two high altitude species *A. thomsoni* and *A. tibetana* from the Northwest Himalaya occur in the Pamirs and the Central Tien Shan also. *Cynoglossum* occurs in subtropical and temperate mountains, the plains of north India, the Himalaya and the Central Tien Shan and the Eastern and Western Ghats and mountains of Malaya. *Echinospermum* extends from the Ural region across the Altai Mountains, Middle Asia, the Hindu Kush and the Himalaya and some species are known from North America also. *E. barbatum* from the Northwest Himalaya is known from Afghanistan, Middle Asia, Pamirs, Central Tien Shan, the Altai Mountains and the Urals. *Eritrichium* is known from Europe and the Middle Asiatic mountains. *Lindelofia* seems to be restricted to the Pamirs, Central Tien Shan and the Himalaya. *Macrotomia* extends from Siberia over the Central Tien Shan and Pamirs to Caucasus. *Microula* is apparently restricted to the Northwest Himalaya. *Myosotis* occurs in Europe, North Africa, Siberia, North America, Middle Asia and Mt. Kenya in East Africa. *M. caespitosa* from the Northwest Himalaya occurs in Afghanistan, Middle Asia, Siberia, Europe, North Africa and North America. *Onosma* extends from the Mediterranean steppes eastwards to Middle Asia. *Paracaryum* occurs in southern Europe, western and Middle Asia and the Western Ghats. *Rochelia* occurs in southern Europe, North Africa and extends eastwards to Songoria, the Hindu Kush and the Pamirs. *Solenanthus* extends from southern Europe through western Asia across the Hindu Kush to Middle Asia, Songoria and the Altai Mountains. *S. circinatus* from the Northwest Himalaya is known also from Afghanistan, the Altai Mountains, Songoria, southern Europe and the Central Tien Shan and Pamirs.

SCROPHULARIACEAE

Lagotis is known from North Asia and North America and from the Middle Asiatic mountains. *L. glauca* from the Northwest Himalaya is reported from North Asia and North America. The genus *Pedicularis* is widely distributed in arctic-alpine areas and is circumpolar in Europe, North Asia, North America, Siberia and occurs in Middle Asia, Central Tien Shan, Pamirs, the Himalaya, Hindu Kush, Iran, Songoria and the Western Ghats. Some of the high altitude species of genus from Northwest Himalaya extend westwards to Europe as far as the Pyrenes and northeastwards to North America. *P. verticillata* occurs, for example, in Europe, arctic North Asia and North America. *P. versicolor* occurs in Europe, arctic North Asia and North America. *Picrorhiza* is apparently restricted to the Himalaya. *Scrophularia* is distributed in north temperate of Europe, western and Middle Asia, Caucasus, Hindu Kush and the Himalaya. *S. variegata* from the Northwest Himalaya extends to the Caucasus. *Veronica* is known from Europe, North and South Africa, North America, western Asia, Caucasus, Hindu Kush, Middle Asia, the Himalaya, Khasi Hills and the Western Ghats. *V. anagallis* from the Northwest Himalaya is also known from Europe, North and South Africa, North America, the Khasi Hills and the Western Ghats. *V. ciliata* occurs in Middle Asia and Siberia and *V. serpyl-*

lifolia occurs in Europe, North Asia, North and South America and North Africa.

LABIATAE

Calamintha is known from Europe, North Africa, western Asia, Caucasus, northern Asia, Middle Asia, Hindu Kush, the Himalaya, Khasi Hills, Western Ghats, the mountains of Sri Lanka, Mt. Kenya and Canada. *C. chenopodium*, found occasionally in the lower reaches of the high altitude ecosystem of the Northwest Himalaya, is known also from the Khasi Hills, the Western Ghats and Sri Lanka and extends westwards to Afghanistan and the Caucasus. *Dracocephalum* extends from southern Europe eastwards to Songoria, Dahuria, Siberia and Middle Asia. *Elsholtzia* seems to be restricted to tropical mountains like the hills of Malaya and Burma, the Khasi Hills and the tropical Himalaya. *Hyssopus* is distributed in eastern Europe and western Asia. *Marrubium* is a genus of the temperate lands and is known from Europe, North Africa and the Altai Mountains. *M. lanatum* from the Northwest Himalaya occurs also in the Altai Mountains. *Nepeta* is distributed in temperate Europe, North Africa, Caucasus, Middle Asia, Songoria and the Western Ghats. *Origanum* is distributed from Europe and North Africa through western Asia, Middle Asia and the Himalaya. *O. vulgare* is widely distributed in these areas. *Perovskia* is apparently restricted to Middle Asia. *Phlomis* is recorded from the Philippines, Java, Tien Shan, Pamirs and the Himalaya. *Scutellaria* is distributed in the north temperate zone and species are reported from the Andes also; it is besides known from the Khasi Hills and the Western Ghats and the mountains of Sri Lanka in addition to the Middle Asiatic mountains. *Thymus* is widely distributed in Europe, western Asia, northern and Middle Asia, North Africa and the Himalaya.

CHENOPODIACEAE

The genus *Axyris* has an interesting distribution from the northwest America, North China, Siberia, the Altai Mountains, Songoria, Central Tien Shan, Pamirs, western Tibet and the Himalaya. *A. amaranthoides*, occurring in the high altitude ecosystem of the Northwest Himalaya, is known also from western Tibet, Middle Asia, the Altai Mountains, Songoria, North China and Siberia. *Chenolea* is known from Europe, North Africa, Songoria, Middle Asia, western Tibet and North America. The high altitude species *C. divaricata* from the Northwest Himalaya is distributed in Songoria, the Alai-Pamirs and the Central Tien Shan. *Corispermum* is known from southern Europe through Caucasus, Middle Asia, China, Songoria and northwest America. *C. hyssopifolium* from the Northwest

Himalaya is distributed outside in southern Europe, Caucasus and China. *Eurotia* is distributed in Europe, Siberia Mongolia, northwest America and Middle Asia. The widely distributed *E. ceratoides* occurs in the Northwest Himalaya. The genus *Kochia* extends from southern Europe and North Africa to Middle Asia, northwest America and the Western Ghats. *Haloxylon* is known from southern Europe, West and Middle Asia, China, the Hindu Kush, Burma and northwest India. *Microgynoecium* seems to be restricted to the Himalaya. *Salsola* is known from Europe, North Africa, North Asia, Siberia, North America and Middle Asia. The genera *Atriplex* and *Chenopodium* are basically Mediterranean steppes plants.

POLYGONACEAE

Oxyria is found on the mountains of Europe, in northern Asia, Middle Asia and North America. *Polygonum* has a circumpolar and arctic-alpine distribution and is known from Europe, Asia and North America. *Rheum* seems to be confined to Middle Asia and the Himalaya. *Rumex* is distributed in Europe, western Asia, Caucasus, North Asia, North America, North Africa, the Himalaya, Middle Asia, the Khasi Hills and the Western Ghats.

ELAEAGNACEAE

The genus *Hippophaë* occurs in Middle Asia and the Hindu Kush; *H. rhamnoides* is the species that is common to the Northwest Himalaya, Hindu Kush, Pamirs and the Central Tien Shan.

THYMELEACEAE

Stellera occurs in Caucasus, Middle Asia, Pamirs and the Northwest Himalaya.

MONOCOTYLEDONS

Eremurus is known from Middle Asia, the Central Tien Shan, Pamirs and the Northwest Himalaya. *Fritillaria* occurs in Middle Asia and the Northwest Himalaya. *Allium* is widely distributed in Europe, northern Asia, Northwest America, the Altai Mountains and Middle Asia. Some of the high altitude species from the Northwest Himalaya occur widely outside. *A. victorialis* occurs, for example, in Europe, North Asia and northwestern America; *A. schrenkii* occurs in Siberia; *A. odorum* in North Asia and *A. atropurpureum* in Europe, Siberia. *Lloydia* occurs in the Alps, North and Middle Asia, North America and the Central Tien Shan and Pamirs. *Gagea* is known from Europe, North Asia, Middle Asia, Central Tien Shan and

Pamirs. *Trillium* seems to be confined to the Himalaya. *Polygonatum* is known from Europe, North Asia and Middle Asia. *Juncus* occurs in Europe, North Asia, Middle Asia, the Hindu Kush, Himalaya and the Khasi Hills. *Cobresia* is distributed in the Himalaya, the Hindu Kush and the Middle Asiatic mountains. *Carex* is known from Mt. Kilimanjaro, Pamirs, Central Tien Shan, the Himalaya, Khasi Hills and the Western Ghats. *Scirpus* occurs in the arctic north, Middle Asia, the Himalaya, Khasi Hills, the Western Ghats and in the hills of Sri Lanka. *Stipa* is known from Korea, Mongolia, the Altai Mountains, Siberia, Central Tien Shan, Pamirs, Iran, Hindu Kush and the Himalaya. *Stipa orientalis* from the Northwest Himalaya extends to the Altai Mountains; *S. mongolica* to the Altai Mountains and Mongolia and *S. sibirica* to Afghanistan, Siberia and Korea. *Calamagrostis* is known from western Asia, Europe, Africa and the Middle Asiatic mountains. *C. epigejos* from the Northwest Himalaya occurs also in Europe, northern Asia and Africa. *Deyeuxia* occurs in Europe, Caucasus, Middle Asia, the Himalaya and the Khasi Hills. *Deschampsia*, a widespread arctic-alpine genus, is known also from Mt. Kilimanjaro and Middle Asia. *Poa* is widely distributed in Europe, North and Middle Asia, the Altai Mountains, Central Tien Shan, Pamirs, the Himalaya and the Khasi Hills. *P. nemoralis* from the Northwest Himalaya is also known from Europe, North Asia and North America. *Festuca* occurs in Central Europe, western and Middle Asia, Mt. Kilimanjaro and Mt. Kenya in East Africa, Caucasus, Dahuria, Siberia, the Himalaya and North America. *F. valeriana* from the Northwest Himalaya is known also from southern and Central Europe, Middle Asia and North America. *F. altaica*, another Northwest Himalayan species, is distributed in the Altai Mountains, Siberia and Middle Asia and *F. sibirica*, also from the Northwest Himalaya, occurs in Caucasus, Dahuria, the Altai Mountains and Siberia. The genus *Bromus* is known from Europe, North, West and Middle Asia and the Himalaya. The high altitude species *B. patulus*, from the Northwest Himalaya, is known also from Europe and western and northern Asia. *Glyceria* is distributed in Europe and in northern Asia. The genus *Elymus* is distributed in Abyssinia, Afghanistan, Middle Asia, Siberia and North America.

3. FLORISTIC COMPONENTS AND AFFINITIES

1. Components

We may now consider the major floristic components of the high altitude ecosystem of the Northwest Himalaya. Although it is customary to differentiate the geographical components from elements, which are plants grouped in accordance with the region reflecting the genesis of the flora [114], in the present discussion we employ the term in a broad sense. It is not possible, in the present state of our knowledge, to accurately separate the elements from components. On the basis of the material brought together in earlier pages, we might conclude that broadly speaking, the high altitude plants of the Northwest Himalaya belong to four major floristic components, 1) the boreo-alpine (arctic-alpine) and specially the circumpolar forms, 2) the Euro-Siberian or the Trans-Palaearctic forms, 3) the southern European or the Mediterranean forms and 4) the Siberian-Mongolian (Manchurian) forms. These major components are on the whole nearly equally strongly developed in the high altitude ecosystem of the Northwest Himalaya. Their percentage frequency is diagrammatically represented in Fig. 45.

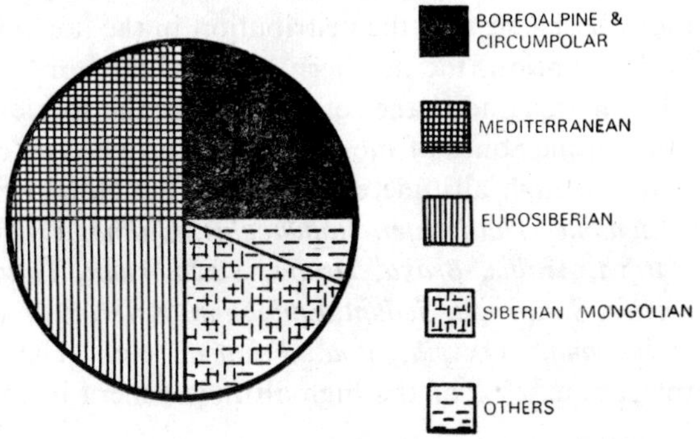

Fig. 45. The percentage frequency of the major phytogeographical components in the high altitude ecosystem of the Northwest Himalaya.

As is well known, the typical boreo-alpine species are more or less conspicuously discontinuously distributed in the arctic north and on the so-called alpine zones of mountains in the south. Their total range of distribution has therefore two major disjunct areas viz., a northern and a southern area, separated by a more or less wide space in which they are absent. The northern area is generally lowland and the southern area is mountainous. Boreo-alpine distribution is striking and rather commonly met with among many groups of animals (especially insects) and plants in Europe, but is less frequent in Asia. The relative poverty of true boreo-alpine species in Asia is closely correlated to the well known fact that the Pleistocene glaciation and ice-sheet cover were much less extensive in Asia than either in Europe or North America [3]*. The boreo-alpine distributional pattern is indeed generally explained as a part of the effects of Pleistocene glaciations and the boreo-alpine species are also known as Pleistocene relicts. The plants, which we include here under the boreo-alpine

*Yet the boreo-alpine plants in Asia are perhaps larger than in America.

group, may not often all have a typically discontinuous distribution. They are nevertheless distributed in the boreal arctic and southern mountainous areas and their distribution pattern is perhaps in some way related to the Pleistocene glaciations [17]. We apply the term boreo-alpine, therefore, in a general sense so as to include all the plants, which occur in the cold, arctic boreal areas and at high altitudes on mountains in a southern area (north of the Equator), irrespective of the fact whether they occur in the intervening lowlands or are absent. The boreo-alpine distribution may be strictly regional and restricted to the arctic north of Europe and the southern area also on European mountains, although some of the species may occasionally extend eastwards to the Himalaya, where they find their extreme eastern limits. Their northern area also often extends eastwards to North Asia and Siberia, so that the distribution in the north is Eurasiatic. In circumpolar distribution, the northern range of the boreo-alpine forms extend to North America and in the southern range the species are as usual distributed in the alpine zones of mountains in Europe-Asia. The circumpolar components in high altitude ecosystem of the Northwest Himalaya are species of *Anemone*, *Thalictrum*, *Ranunculus*, *Caltha*, *Aquilegia*, *Aconitum*, *Actaea*, *Parrya*, *Arabis*, *Braya*, *Draba*, *Sisymbrium*, *Thlaspi*, *Lychnis*, *Stellaria*, *Potentilla*, *Saxifraga*, *Sedum*, *Epilobium*, *Aster*, *Primula*, *Gentiana*, *Pedicularis*, *Polygonum*, *Oxyria*, *Poa*, *Festuca*, etc. The circumpolar plants constitute about 25% of the high altitude genera in the Northwest Himalaya.

Particular attention may be drawn here to an interesting peculiarity of the boreo-alpine component. While the boreo-alpine insects are relatively quite unimportant elements in the high altitude ecosystem of the Northwest Himalaya [76], the plants with boreo-alpine pattern of distribution are among the dominant forms. Unlike the boreo-alpine insects, these plants are not in the nature of Pleistocene relicts in the Northwest Himalaya. The fact of their abundance at altitudes of 3900–5000 m would seem to indicate their having differentiated in Pre-Pleistocene times in the Northwest Himalaya.

The second major component is the Euro-Siberian group of genera, which are widely distributed in Asia and Europe, where they extend sometimes to the extreme north. They are generally found in the Alps, Carpathians and Central Europe and extend mostly eastwards across the Caucasus, the Hindu Kush and the Himalaya to the Pamirs, Central Tien Shan, the Altai Mountains and sometimes even to North America. In some cases the species found in the Northwest Himalaya are completely identical with those occurring in the Alps, Siberia and the Altai Mountains. The common examples of this component are *Adonis*, *Callianthemum*, *Cimifuga*, *Isopyrum*, *Stylophorum*, *Meconopsis*, *Alyssum*, *Crambe*, *Viola*, *Cerastium*, *Trollius*, *Geranium*, *Gypsophila*, *Silene*, *Oxytropis*, *Geum*, *Bupleurum*, *Pimpi-

nella, *Lonicera*, *Valeriana*, *Leontopodium*, *Inula*, *Crepis*, *Hieracium*, *Diapensia*, *Androsace*, *Lysimachia*, *Scrophularia*, *Eurotia*, *Allium*, *Lloydia*, *Gagea*, *Polygonatum*, etc. This component is basically a Trans-Palearctic group and is nearly as numerous as the boreo-alpine-circumpolar group in the high altitude ecosystem of the Northwest Himalaya. There is no doubt that these two groups are closely interrelated and represent perhaps members of the same larger single complex.

The third major component viz., the Southern European or the Mediterranean forms is nearly as numerous as any of the other two which we have considered before. The genera belonging to this group are distributed primarily in southern Europe, North Africa, western Asia and extend eastwards to Afghanistan, the Himalaya and the Alai-Pamirs. The common examples of this group are species of *Delphinium*, *Hypecoum*, *Cheiranthus*, *Barbarea*, *Alyssum*, *Erysimum*, *Capsella*, *Iberidella*, *Arenaria*, *Myricaria*, *Ononis*, *Trifolium*, *Heracleum*, *Scabiosa*, *Artemisia*, *Senecio*, *Arctium*, *Echinops*, *Carduus*, *Acantholimon*, *Swertia*, *Myosotis*, *Rochelia*, *Arnebia*, *Onosma*, *Origanum*, *Thymus*, *Calamintha*, *Nepeta*, *Dracocephalum*, *Corispermum*, *Kochia*, etc. Occasionally some species of this group extend to North America and many have of course penetrated Siberia. The species occupy a dominant place in the vegetation of the high altitudes of the Northwest Himalaya. The Mediterranean component is conspicuous within the forest zone and even forms an important part of the lowland flora of northwestern India, Pakistan and Baluchistan [29, 81, 83, 85, 86].

The fourth major group of genera like *Oxygraphis*, *Chorispora*, *Eutrema*, *Biebersteinia*, *Silene*, *Thylacospermum*, *Thermopsis*, *Caragana*, *Astragalus*, *Pleurospermum*, *Allardia*, *Tanacetum*, *Cousinia*, *Saussurea*, *Taraxacum*, *Scorozonera*, *Solenanthus*, *Echinospermum*, *Lagotis*, *Perovskia*, *Axyris*, *Eremurus*, *Fritillaria*, *Cobresia*, etc., are distributed from Siberia, Mongolia, the Altai Mountains, Central Tien Shan, the Alai-Pamirs and the Northwest Himalaya. Some species extend westwards to the Hindu Kush. This group is somewhat smaller than the other three mentioned above, but many species are ecologically dominant in large areas of the Northwest Himalaya.

In addition to these major groups, there is yet another minor floristic group of genera like *Christolea*, *Cremanthodium*, *Jaeshkea*, *Microula*, *Picrorhiza*, *Microgynoecium*, *Trillium*, *Danthonia*, etc., which are apparently confined to the Himalaya and only to a small extent to the Himalaya-Pamirs region.

In striking contrast to the high altitude insect life of the Northwest Himalaya, the high altitude plants are conspicuously poor in endemic genera and species. Most genera and indeed even a large number of the species are common to the larger elevated area, embracing the Alai-Pamirs, western Tibet and in part the Central Tien Shan (*vide infra*).

The grouping which we have recognized here differs somewhat from that suggested by other earlier authors. Hooker [59] recognized, for example, the European, Siberian and Malayan groups of plants in the Himalaya. Clarke [20] recognized the Indo-African component in the arid western end of the Himalaya, a so-called Central Asia (actually the Middle Asiatic) component of Tertiary age and finally a European component that spreads eastward along the Himalaya, after the Himalayan System had already been uplifted to elevations above 4000 m.

The European component of these authors includes the boreo-alpine-circumpolar and a part of the southern European and Euro-Siberian components described here. The Indo-African component corresponds generally in part to the Mediterranean group. The so-called Central Asiatic component of earlier authors corresponds to the Mongolian-Siberian-Middle Asiatic group.

2. *Affinities*

There is considerable prevailing confusion regarding the phytogeographical affinities of the Himalaya. Most authors generally place the Himalaya under the Sino-Himalayan-Tibetan Mountain Province of the Sino-Japanese Region of the Boreal Kingdom of Good [47]. It seems rather strange that in this arrangement, botanists like Gaussen [42], Good [47] and others consider the Himalaya more as a sort of phytogeographical "barrier" than in any other floristic-phytogeographical sense. Other recent workers like Banerji [7] and Chatterjee [18] have, however, correctly recognized the basic difference between the forested slopes and the elevated regions higher above. They consider the southern slopes of the Himalaya as comprising two separate Provinces viz., the Eastern Himalayan Province and the Western Himalayan Province.

The boundary between these two Himalayan "Provinces" is generally assumed by most workers to be sharply defined and is placed at 85° EL, between western Nepal and the eastern parts of the Kumaon Himalaya. We find, however, that there are two-fold objections to this idea. It has been amply demonstrated by us in earlier publications [76, 78] that the biogeographical character of the Himalaya presents a gradual transition from the east to the west, in which a sharply defined boundary cannot be recognized. The Eastern phytogeographical section of the Himalaya passes into the Western section imperceptibly. The eastern flora and faunas tend to lose their dominance, as we pass westwards, and come to be gradually replaced by western elements, which correspondingly gain in dominance towards the west. The evidence of distribution of plants, insects and other animals, geological history, geographical peculiarities, climate and a variety of other ecological characters show conclusively that this transition from the eastern to the definitive western flora must be placed much further west than has been supposed by other authors. We place this transition in the great defile of the R. Sutlej (*vide* Chapter I). The Sutlej defile is an abrupt physical break, but forms a transitional area ecologically and biogeographically.

According to the estimates of competent authorities, about three thousand species or almost 29% of the total species of Dicotyledons from India and about a thousand species of Monocotyledons are believed to be endemic in the Himalayan region. These estimates refer to the Himalaya as a whole and to its entire flora, including the forested slopes and high altitude areas above the timberline. The Eastern Himalaya, together with the mountains of Assam, Yunnan, North Burma and parts of northern Indo-China are believed by several botanists like Takhtajan [113] to have been the centre of origin and radiation of the whole of the temperate flora of the northern hemisphere. They do not consider the region to be a refugium of the relict floras. It is believed further that in this region occur the phylogenetic series that link the temperate forms with the subtropical and tropical forms.

While these ideas may have perhaps a large measure of truth with reference to the Eastern and Central Himalaya and in particular for the vegetation below the timberline, they need considerable modification in case of the high altitude plants of the Northwest Himalaya.

Some attention may now be directed to certain weak points in existing theories. The earlier ideas on the phytogeographical divisions and affinities of the Himalaya would seem to imply the utterly illogical supposition that the entire Himalaya, from its extreme eastern end to the extreme western end, arose about the same time. The geological and other evidence has shown, however, that the uplift of the Northwest Himalaya is a later event than that of the rest of the Himalaya. It would follow from this that the so-called Western Himalayan Province of earlier authors is in reality much younger than the Eastern Himalayan Province; the two have had wholly different histories and contributory sources.

Current ideas also apparently fail to take into account the striking ecological and floristic differences between the Northwest Himalaya and the rest of the Himalaya. They also totally ignore the distinctiveness of the ecosystem above the timberline. The Eastern Himalayan Province, as recognized by many earlier workers, may perhaps be valid in part and its affinities may even lie with the Sino-Japanese Region. We are not, however, specifically concerned here with this part of the Himalaya. The Western Himalayan Province of these authors has certainly no close affinities with the Sino-Japanese Region. It is also not correct to treat the area as major Province, distinct and isolated from the vast elevated land further to the north of Kashmir. It is not indeed a major phytogeographic division, but only a part of such a large division, the bulk of which lies wholly outside the geographical limits of India.

We have to consider the so-called Western Himalayan Province in the light of the brilliant researches of Russian botanists like Fedchenko [31–35], Korovin [67–69], Paulsen [91], Popov [92–93], Raikova [95], Stanyukovitch [109] and others. Their work in Middle Asia has thrown considerable new light on the phytogeographical peculiarities of the elevated regions comprising the Northwest Himalaya, the Alai-Pamirs and the Central Tien Shan. From the evidence painstakingly assembled together by these workers, it is concluded that the Northwest Himalaya is an integral part of the larger floristic area of Middle Asia, rather than of the Sino-Japanese area.

In an earlier chapter we have demonstrated and emphasized the fundamental similarity of the general ecology and vegetation and the plant formations and associations between the Northwest Himalaya and the Pamirs. It was also suggested that, ecologically speaking, the Northwest Himalaya is the extreme southern fringe zone of the vegetation of Middle Asiatic mountains. Not only is there a remarkable affinity between the Northwest Himalaya and the Pamirs, but a similar basic affinity to the Central Tien Shan may also be detected. We find, for example, that about 85% of the genera and nearly 70% of the species of high altitude plants from the Northwest Himalaya are identical with those of and common to the Pamirs. It is also interesting to observe that 75% of the genera and nearly 60% of the species are likewise common to the Central Tien Shan.

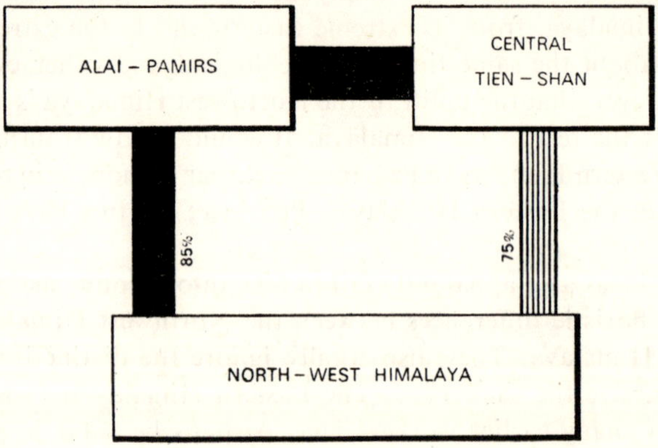

Fig. 46. The phytogeographical affinities of the Northwest Himalaya, the Alai-Pamirs and the Central Tien Shan Highland region of Middle Asia. The figures indicate the mean percentage of elements which are common to the Northwest Himalaya and the Pamir on one side and the Northwest Himalaya and the Central Tien Shan on the other side. There is greater affinity to the Pamir than to the Tien Shan.

Figure 46 represents diagrammatically the close floristic affinities between the Northwest Himalaya, Pamirs and the Central Tien Shan. Most of the genera, which are common to the Central Tien Shan and to the Northwest Himalaya, occur naturally in the Pamirs. Over 155 genera of the high altitude plants from the Northwest Himalaya are common to the Pamirs and an equal number to the Central Tien Shan. It is found that in a great many genera even the species are wholly identical. Even more remarkable is the fact that the same species compose nearly the same ecological formations and plant associations or at least similar formations in the Northwest Himalaya and the Pamirs. The few genera found in the Northwest Himalaya, but which are absent in the Pamirs, are known to be distributed either in western Asia, North Africa and southern Europe or they are in the nature of outliers of the Eastern Himalayan flora. A small minority may perhaps be endemic in the Northwest Himalaya. We may now pay some attention to the genera common to the Northwest Himalaya and the Middle Asiatic mountains by way of illustration of the foregoing conclusions.

3. Plants Common to the Pamirs

The examples of plants common to the high altitude ecosystem of the Northwest Himalaya and the Pamirs include the following:

RANUNCULACEAE

Aconitum napellus, A. rotundifolium, Adonis aestivalis, A. chrysocyathus, Anemone biflora, A. obtusiloba, Aquilegia, Callianthemum, Delphinium elatum, Oxygraphis glacialis, Ranunculus, Thalictrum alpinum, T. minus and *Trollius.*

FUMARIACEAE

Hypecoum, Corydalis gortschavakovi and *C. govanianum.*

CRUCIFERAE

Alyssum minimum, Arabis, Barbarea, Braya, Capsella procumbens, Christolea crassifolia, Chorispora elegans, C. sibirica, C. tenella, Crambe kotschyana, Draba alpina, D. altaica, D. stenocarpa, D. tibetica, D. verna, Eruca sativa, Erysimum altaicum, E. hieracifolium, Eutrema, Parrya excapa, Sisymbrium and *Thlaspi arvense.*

VIOLACEAE

Viola.

Caryophyllaceae

Acanthophyllum, *Arenaria griffithii*, *A. serpyllifolia*, *Cerastium dahuricum*, *Gypsophila*, *Silene*, *Stellaria umbellata*, *Thylacospermum caespitosum*, etc.

Tamariscaceae

Myricaria.

Geraniaceae

Biebersteinia, *Erodium cicutarium*, *Geranium collinum* and *G. rectum*.

Leguminosae

Astragalus alpinus, *A. densiflorus*, *A. tibetanus*, *A. tribuloides*, *Caragana pygmaea*, *Hedysarum cephalotes*, *Lotus corniculatus*, *Medicago falcata*, *M. minima*, *Melilotus albus*, *M. officinalis*, *Oxytropis glabra*, *Thermopsis*, *Trifolium pratense*, *Trigonella emodi* and *T. geminiflora*.

Rosaceae

Cotoneaster multiflora, *C. mummularia*, *Fragaria vesca*, *Geum urbanum*, *Potentilla biflora*, *P. gelida*, *P. hololeuca*, *P. moorcroftii*, *P. nivea*, *P. repens*, *P. sericea*, *Rosa pimpinellifolia*, *R. webbiana*, *Spiraea*, *Sibbaldia procumbens*, *S. tetrandra*, etc.

Rhamnaceae

Rhamnus catharticus.

Saxifragaceae

Parnassia laxmanii, *Saxifraga cernua*, *S. flagellaris*, *S. hirculus*, *S. oppositifolia* and *S. setigera*.

Crassulaceae

Sedum coccineum and *S. ewersii*.

Onagraceae

Epilobium hirsutum and *E. palustre*.

Plumbaginaceae

Acantholimon.

Primulaceae

Androsace lehmanniana, A. villosa, Primula kaufmanniana, P. moorcroftiana and *P. sibirica.*

Gentianaceae

Gentiana aurea, G. decumbens, G. humilis, G. olivieri, G. prostrata, Pleurogyne caranthiaca and *Swertia.*

Umbelliferae

Bupleurum, Eryngium, Heracleum and *Pleurospermum.*

Caprifoliaceae

Lonicera heterophylla, L. hispida and *L. microphylla.*

Valerianaceae

Valeriana.

Dipsacaceae

Dipsacus, Morina and *Scabiosa.*

Compositae

Achillea, Anaphalis, Artemisia annua, A. desertorum, A. dracunculus, A. lehmanniana, A. macrocephala, A. persica, A. pontica and *A. sacrorum, Aster alpinus, Carduus, Centaurea, Cirsium, Cousinia microcarpa, Crepis multicaulis, C. sibirica, Doronicum, Echinops, Erigeron, Helichrysum, Hieracium umbellatum, H. virosum, Inula helenium, I. oculus-christi, I. rhizocephala, Jurinea, Lactuca orientalis, L. scariola, Leontopodium alpinum, Ligularia, Saussurea sorocephala, Scorzonera, Senecio coronopifolius, S. jacobaea, S. subdentatus, S. thomsonii, Sonchus arvensis, Tanacetum* and *Taraxacum.*

Boraginaceae

Arnebia, Echinospermum, Eritrichium villosum, Lindelofia anchuosides,

Macrotomia euchroma, *Myosotis*, *Onosma*, *Solenanthus circinatus* and some other species and *Rochelia*.

SCROPHULARIACEAE

Lagotis glauca, *Pedicularis pycnantha*, *P. rhinanthoides*, *Scrophularia*, *Verbascum*, *Veronica anagalis*, *V. becaebunga*, *V. campylopoda* and *V. ciliata*.

LABIATAE

Dracocephalum nutans, *D. stamineum*, *Nepeta longibracteata*, *Origanum vulgare*, *Perovskia abratanoides*, *Phlomis oreophila*, *Scutellaria* and *Thymus*.

CHENOPODIACEAE

Atriplex, *Chenopodium album*, *C. botrys*, *Corispermum*, *Eurotia ceratoides*, *Kochia prostrata* and *Salsola*.

POLYGONACEAE

Calligonum, *Oxyria digyna*, *Polygonum*, *alpinum*, *P. amphibium*, *P. aviculare*, *P. rupestre*, *P. viviparum*, *Rheum nobile* and *Rumex*.

ELAEAGNACEAE

Hippophaë rhamnoides.

GNETACEAE

Ephedra ciliata, *E. distachya*, *E. glauca* and *E. intermedia*.

IRIDEAE

Iris.

LILIACEAE

Allium fedchenkoanum, *A. griffithianum*, *A. pallasii*, *A. platyspathum*, *Eremurus*, *Fritillaria*, *Gagea reticulata*, *Lloydia serotina* and *Polygonatum*.

CYPERACEAE

Carex acutiformis, *C. divisa*, *C. griffithii*, *C. infuscata*, *C. melanantha*,

C. nivalis, C. orbicularis and *C. parva, Cobresia capillifolia, C. filifolia, C. royleana, C. schoenoides* and *C. stenocarpa, Scirpus compressus, S. lacustris, S. littoralis, S. maritimus, S. pauciflorus, S. tabernemontanus* and *S. triquetrus.*

GRAMINEAE

Bromus danthoniae, B. inermis, B. macrostachys, B. oxyodon, B. ramosus, B. scoparius and *B. tectorum, Calamagrostis emodensis, C. epigejos, Deschampsia caespitosa, D. koelerioides, Phragmites communis, P. isiaca, Poa alpina, P. attenuata, P. bulbosa vivipara, P. nemoralis* and *P. pratensis, Ptilagrostis mongolica, Stipa breviflora, S. pulcherrima* and *S. sareptana.*

4. Plants Common to the Central Tien Shan

The examples of plants common to the high altitude ecosystem of the Northwest Himalaya and the Central Tien Shan comprise the following [43, 44, 45 & 46]:

RANUNCULACEAE

Aconitum, Adonis, Aquilegia, Delphinium, Callianthemum, Isopyrum, Oxygraphis glacialis, Ranunculus aequatilis, R. affinis, R. longicaulis, R. natans, R. pedatifidus, R. pulchellus and *R. repens, Thalictrum* and *Trollius.*

PAPAVERACEAE

Hypecoum.

FUMARIACEAE

Corydalis.

CRUCIFERAE

Alyssum, Arabis, Barbarea, Brassica, Braya, Capsella bursa-pastoris, Chorispora sibirica, C. tenella, Christolea crassifolia, Draba altaica, D. fladnizensis, D. lasiophylla, D. stenocarpa and *D. tibetica, Erysimum, Eutrema, Lepidium, Parrya excapa, Sisymbrium, Smelowskia* and *Thlaspi.*

VIOLACEAE

Viola.

Caryophyllaceae

Arenaria griffithii, *A. meyeri*, *A. serpyllifolia*, *Cerastium cerastoides*, *C. dahuricum*, *Dianthus barbatum*, *D. caryophyllus*, *D. chinensis*, *Gypsophila*, *Hermiaria hirsuta*, *Sagina*, *Saponaria*, *Silene conoides*, *S. graminifolia*, *S. repens*, *S. wallichiana*, *Stellaria graminea* and *S. media*.

Geraniaceae

Erodium cicutarium, *Geranium collinum*, *G. pratense* and *G. rectum*.

Leguminosae

Astragalus alpinus, *A. densiflorus*, *A. macropterus*, *A. ophiocarpus*, *A. tibetanus*, *Caragana*, *Hedysarum cephalotes*, *Lotus*, *Medicago*, *Melilotus alba*, *Onobrychis*, *Oxytropis glabra*, *O. lapponica*, *Thermopsis*, *Trifolium fragiferum*, *T. pratense*, *T. repens*, *Trigonella*, etc.

Rosaceae

Agrimonia, *Alchemilla*, *Cotoneaster multiflora*, *Fragaria*, *Geum*, *Potentilla biflora*, *P. desertorum*, *P. gelida*, *P. hololeuca*, *P. moorcroftii*, *P. nivea*, *P. repens* and *P. sericea*, *Sanguisorbus*, *Sibbaldia* and *Sorbus*.

Saxifragaceae

Parnassia laxmanii, *P. palustris*, *Saxifraga cernua*, *S. flagellaris*, *S. hirculus*, *S. oppositifolia*, *S. setigera* and *S. sibirica*.

Crassulaceae

Sedum ewersii.

Onagraceae

Epilobium hirsutum and *E. palustre*.

Plumbaginaceae

Acantholimon.

Primulaceae

Androsace septentrionalis, *Lysimachia* and *Primula sibirica*.

Gentianaceae

Gentiana azura, *G. barbata*, *G. riparia*, *G. tenella* and *Swertia*.

Umbelliferae

Bupleurum, *Eryngium*, *Heracleum* and *Pimpinella*.

Caprifoliaceae

Lonicera hispida and *L. microphylla*.

Dipsacaceae

Dipsacus, *Morina* and *Scabiosa olivieri*.

Compositae

Achillea millefolium, *Artemisia annua*, *A. absinthium*, *A. lehmanniana*, *A. macrocephala*, *A. persica*, *A. scoparia* and *A. viridis*, *Aster alpinus* and *A. altaicus*, *Carduus*, *Cirsium*, *Cousinia*, *Crepis multicaulis*, *Doronicum*, *Echinops*, *Erigeron alpinus*, *Helichrysum*, *Inula rhizocephala*, *Jurinea*, *Lactuca*, *Leontopodium alpinum*, *Ligularia*, *Saussurea sorocephala*, *Scorzonera*, *Senecio*, *Sonchus*, *Tanacetum tenuifolium* and *Taraxacum*.

Boraginaceae

Arnebia tibetana, *Cynoglossum*, *Echium*, *Eritrichium villosum*, *Lindelofia*, *Lithospermum arvense*, *L. officinale*, *L. tenuiflorum*, *Macrotomia euchroma*, *Myosotis caespitosum*, *M. palustris*, *M. sylvatica*, *Onosma* and *Solenanthus*.

Scrophulariaceae

Lagotis decumbens, *Linaria*, *Pedicularis abratanifolia*, *P. alberti*, *P. cheiranthifolia*, *P. pycnantha*, *P. rhinanthoides*, *Verbascum* and *Veronica*.

Labiatae

Dracocephalum bipinnatum, *D. heterophyllum*, *D. nutans*, *D. stamineum*, *Eremostachys*, *Mentha*, *Nepeta cataria*, *Origanum vulgare*, *Phlomis oreophila*, *Scutellaria galericulata* and *Thymus*.

CHENOPODIACEAE

Atriplex, *Axyris amaranthoides*, *A. hybrida*, *Eurotia ceratoides*, *Kochia prostrata* and *Salsola*.

POLYGONACEAE

Oxyria digyna, *Polygonum alpinum*, *P. convolvulus*, *P. hydropiper*, *P. persicaria*, *P. sibiricum*, *P. viviparum* and *Rumex*.

ELAEGNACEAE

Hippophaë rhamnoides.

LILIACEAE

Allium atropurpureum, *A. atrosanguineum*, *A. caeruleum*, *A. oreosprasum*, *A. pallasii*, *A. platyspathum*, *Eremurus*, *Fritillaria*, *Gagea*, *Lloydia serotina*, *Polygonatum roseum* and *Tulipa*.

CYPERACEAE

Carex, *Cobresia stenocarpa*, etc.

GRAMINEAE

Andropogon ischaemum, *Calamagrostis epigejos*, *Digitaria linearia*, *Deschampsia*, *Bromus*, *Festuca*, *Poa*, *Ptilagrostis*, *Stipa breviflora* and *S. capillata*.

4. PHYTOGEOGRAPHICAL DIVISIONS

On the basis of the geographical and physical continuity with the Pamirs, the geological history, general climate, ecology of the high altitude region and the vegetational peculiarities, the floristic and distributional characters, the Northwest Himalaya must be considered as a part of the larger phytogeographic area of Middle Asia. This conclusion is supported by the early suggestion by Clarke [20] and his recognition of the presence of his "Central Asian" component. The phytogeographic divisions of Middle Asia, according to Russian botanists like Fedchenko [31–35], Poppov [92, 93], Korovin [67–69] and others, are both vertical and horizontal, and thus fall into two major areas viz., the lowland desert areas and the high mountain areas. This scheme of division clearly stresses the importance of taking into consideration both the vertical and horizontal distribution.

These major areas are divided into the following Provinces:

1. Turanian Province.
2. Central Kazakhstan Desert Province.
3. South Turkestan Mountain Province.
4. Turkmeno-Hissar Mountain Province.
5. Dhzungaro-Tien Shan Mountain Province.
6. Central Tien Shan High Ranges Province.
7. Pamiro-Tibetan High Ranges Province.

The South Turkestan Mountains and the Pamiro-Tibetan High Ranges Provinces and to a large extent also the Central Tien Shan Province are regions of high altitude, as conceived by us in this book. They are subdivided into a number of Districts as follows:

1. South Turkestan Mountain Province.
   a. Karatau (Calcophyte, lithophyte vegetation).
   b. West Tien Shan (Subsavannah and summer-green vegettaion).
   c. Ferghana (Summer-green and subsavannah).
   d. Kukistan (Juniperus, steppes).
   e. Hissar-Darwaz (Subsavannah, lithophyte, mixed forest).
   f. Badakhshan (Subsavannah, xerophyte shrub, cryophyte).
   g. South Tadjikistan (Juniperus, subsavannah and ephemeric vegetation).
2. Central Tien Shan High Ranges Province.
   a. Southwest District (Steppes, shrubs).
   b. Northwest District (Meadows, steppes).
   c. East Hilly District (Alpine meadows, cushion plants).
   d. Altai District (Steppes, subalpine meadows).
   e. Issyk-kul District (Shrub steppes, meadows, drang conifer forest).
3. Pamiro-Tibetan Highland Province.
   a. Northeast (Cryophyte cushion plants).
   b. Southeast (Alluvial meadows, steppes, semi-desert fragments).
   c. Southwest District (High altitude ecosystem of the Northwest Himalaya).

5. VEGETATION TYPES OF MIDDLE ASIA

According to the researches of Korovin [69], about 30% of the general flora of the Pamirs have elements common with the Central Tien Shan flora. The Pamirs vegetation becomes characteristic at elevations of 3600–4100 m. Three major vegetation types of the Pamirs were recognized by Stannyukovitch in 1949 [109] viz., i) the subalpine, ii) alpine and iii) the nival types at elevations of 6000 m, with cryophyte species like *Oxytropis immersa* and *Sibbaldia tetrandra*. We have already given numerous examples of typical cryophytes vegetation of the Northwest Himalaya, many of which are identical with those described by Russian botanists from areas further north. The characteristic differences between the Central Tien Shan and the Pamirs centre largely around the fact that the former district is arid steppes and the Pamirs cold steppes, cold desert, desert tundra and arctic. The Northwest Himalaya has closer affinities with the Pamirs and may be considered as the Southwest District of the Pamiro-Tibetan Highland Province. Like the Pamirs, the Northwest Himalaya is also cold steppes, cold desert, desert tundra and arctic vegetation.

The principal vegetation types of Middle Asia are grouped by Korovin [69] into the following types:

I. Arid types.
   1. Psammophyte shrub.
   2. Gypsophile subshrub.
   3. Halophyte.
   4. Ephemeric grass.
   5. Valley forest (Tugay).

II. Subarid types.
   6. Mountain steppes.
   7. Mountain subsavannah.
   8. Mountain cryophyte: Xerophyte.
   9. Mountain xerophyte shrub and tree.
   10. Mountain Juniper forest.

III. Humid types.
   11. Mountain summer-green forest and shrub.
   12. Subalpine dark conifer forest (Taiga).
   13. Mountain meadow.

IV. Cryophyte types.
   14. High-mountain (high altitude) cushion plant.

V. Petrophytes.
   15. Lithophyte, chasmophyte.

VI. Hygrophytes.
   16. Aquatic plants.

VII. Eurytypes (Thallophytes).
   17. Lichens.
   18. Moss.

6. CONCLUSIONS

Ecologically and phytogeographically, the Himalaya is a rather complex region of marked east-west and south-north transitions. The major phytogeographical peculiarities of the Himalaya are summarized in the Figs. 47,

48 and 49. Our analysis of available data shows that striking fundamental differences of floristic components and affinities and ecology distinguish the eastern end from the extreme western end of the forest-covered ranges, to the south of Great Himalaya on one hand and the vast elevated areas of the Himalaya *per se*, above the timberline on the other hand. The phytogeographical divisions of the Himalaya into an Eastern Himalayan Province and a Western Himalayan Province by earlier authors is, therefore, an oversimplification. The assumed Sino-Japanese affinity of even their Eastern Himalayan Province, is only valid in parts to the forest-covered mountain ranges to the south, within the limits of India, and to the north, in eastern Tibet, of the Great Himalaya. On the other hand, the extreme high elevations in the Eastern Himalaya are really outside of the Eastern Himalayan Province and also have very little, if any, affinity with the Sino-Japanese

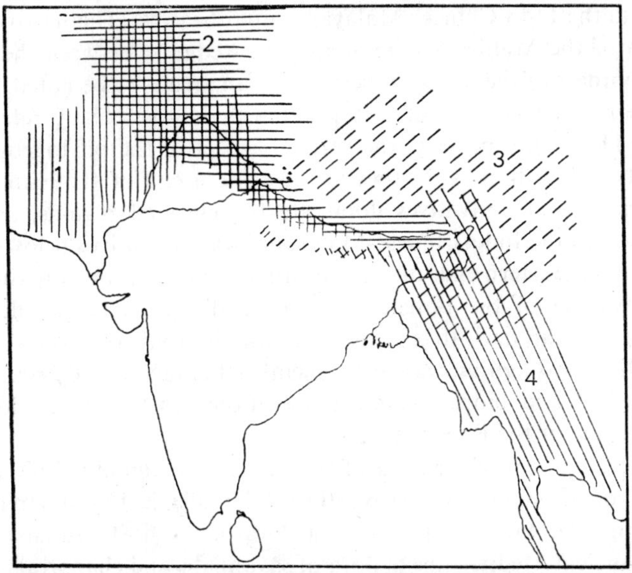

Fig. 47. The major phytogeographical areas of the Tertiary Mountain region. Horizontal aspect of the phytogeography of the high altitude flora of the Northwest Himalaya, showing the principal areas of differentiations of floras. Area 1—is the Mediterranean amphitheatre, 2—the Middle Asiatic amphitheatre (corresponding to the Turkmenian Subregion of the Palaearctic), 3—the Indo-Chinese and 4—the Malayan amphitheatres. In the Northwest Himalaya, Pamirs and parts of Middle Asia, the Mediterranean and Turkmenian amphitheatres overlap and merge into one; there is a narrow tongue-like eastward extension of this complex along the Himalaya, at elevations above the timberline. The Indo-Chinese-Malayan amphitheatres likewise overlap in northern Burma-Assam and extend westward along the wooded slopes of the Himalaya, south of the Great Himalaya and below the timberline in India and also in eastern Tibet north of the Great Himalaya.

floral region. We have here a narrow tongue-like extension of the western floras; indeed a narrow wedge of the Pamiro-Tibetan Province of Middle Asia, constituting a sort of *elevated island* above the tropical-subtropical forest vegetation. In the Northwest Himalaya, the flora of the forest-covered southern mountain ranges is characterized by an admixture of Mediterranean and European floristic components, with some outliers of the eastern floras. The extreme elevations, above the timberline, of the Northwest Himalaya actually mark the southern border or the fringe area of the Pamiro-Tibetan Province. We recognize this area as the Southwest District of the Pamiro-Tibetan Province of Middle Asia and not as a distinct province of the Himalaya. Fig. 47 shows the horizontal aspect of the phytogeographic complexity of the region of Tertiary Mountains in Asia. The principal amphitheatres of the differentiation of floras are the Mediterranean, the Middle Asiatic (corresponding to the Turkmenian Subregion of the Palaearctic) and the Indo-Chinese-Malayan amphitheatres. The narrow eastern extension of the Middle Asiatic areas as a wedge in between the forklike western borders of the Indo-Chinese-Malayan area is to be noted.

A clear understanding of this peculiarity is possible by reference to Fig. 48. This figure presents the vertical aspect of the phytogeographical peculiarities. It will be seen that in the eastern end of the Himalaya the Manchurian floral derivatives dominate the elevated regions, above the timberline and the Indo-Chinese-Malayan floral derivatives below the timberline. In the western end, the high altitude regions, above the timberline, have derivatives of the Boreal-Eurasiatic and Mediterranean floras, but the Mediterranean derivatives extend below the timberline also. In interpreting these two figures, we must remember the fact already stressed viz., the ecological and biogeographical transitions (east-west) are located in the great defile of the R. Sutlej.

The development of the flora of Middle Asia, including also that of the high altitude ecosystem of the Northwest Himalaya, has apparently proceeded from the Arcto-Tertiary forest flora of the northern latitudes and from the ancient Mediterranean flora of savannahs and chaporrals from the south. According to Korovin [69], the sources of the vegetation of this vast elevated region may be traced from the Cretaceous times, through the Palaeocene, Neocene, Quaternary to the Recent times. The formation of dominant desert character, so remarkably conspicuous today, is to be attributed to the Alpine-Himalayan orogenesis. By the time the Himalayan System had been uplifted to elevations of 4000 m above mean sea level, the whole region, together with the Mediterranean and Middle Asia, was well enveloped in aridization, leading to unity of development and likeness on the whole vast territory.

At earlier stages, the gypsophile type of eremophytes came to be differentiated. This type is even at the present time found in the form of a series

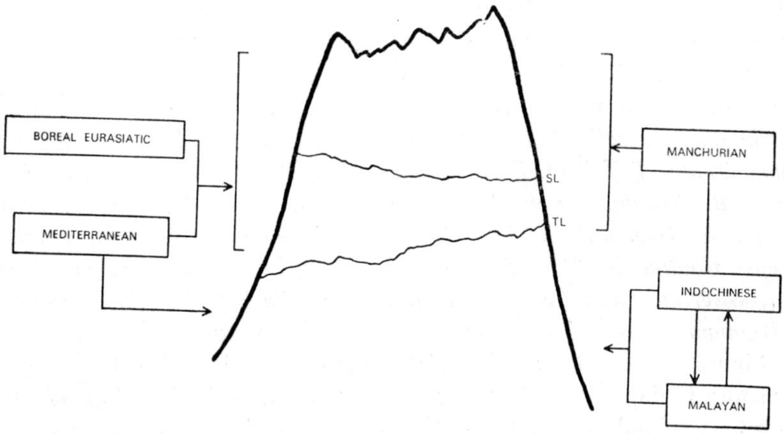

Fig. 48. Vertical aspect of the east-west differences in the phytogeography of the Himalaya. In the eastern end of the Himalaya, the Manchurian floral derivatives are dominant above the timberline and the Indo-Chinese-Malayan complex derivatives below the timberline. In the western end, the high altitude areas, above the timberline, have the derivatives of Boreal Eurasiatic and Mediterranean floras, but below the timberline the Mediterranean derivatives dominate. This figure will have to be taken with Fig. 47 for correct interpretation.

of endemic genera and species from the typical desert families. The formation of mountains merely led to regional differences in the general conditions and also profound changes in the Tertiary flora. The differentiation of the mountain xerophytic and savannah floras is evidently bound up with the main areas of mountain denudation and accumulation.

The Pleistocene Epoch was undoubtedly very important in the history of the oreophile flora and vegetation of the region. It also coincided with the progressive development of nemoral vegetation. A little later the cryophile flora reached its maximum development, due partly to the local transformations *in situ* and partly to undoubted mountain migrations (invasion of elevated regions). The retreat of the Pleistocene glaciers, associated with boreal invasions, naturally resulted in the enormous enrichment of the southern areas, with mountain taiga and mountain meadows. To this factor must also be attributed the presence of the Himalayan plants on the Western Ghats, in the Indian Peninsula. According to Fyson [37], about 12% of the plants on the south Indian Hills are common to the Himalaya; it seems that this number may be higher. Gupta [49] has in fact recently shown that not only the Western Ghats, but also the Eastern Ghats has elements common with the Himalaya. Indeed even as early as 1863, Anderson and later Clarke [20] suggested the Pleistocene migrations of the Himalayan plants to the Peninsula and recognized the important rôle of the Parashnath Hills in this migration [10, 80]. We have recently found some of the typical Himalayan plants in the higher parts of the Satpura and on Mahableshwar also. Common examples of the plants, which are common to the Himalaya and the south Indian mountains, belong to *Anemone*, *Ranunculus*, *Thalictrum*, *Cardamine*, *Capsella*, *Silene*, *Cerastium*, *Arenaria*, *Stellaria*, *Sagina*, *Geranium*, *Oxalis*, *Trifolium*, *Fragaria*, *Potentilla*, *Alchemilla*, *Parnassia*, *Saxifraga*, *Bupleurum*, *Pimpinella*, *Heracleum*, *Lonicera*, *Valeriana*, *Erigeron*, *Anaphalis*, *Gnaphalium*, *Artemisia*, *Senecio*, *Echinops*, *Cnicus*, *Lactuca*, *Crepis*, *Gaultheria*, *Rhododendron*, *Androsace*, *Lysimachia*, *Exacum*, *Gentiana*, *Swertia*, *Cynoglossum*, *Veronica*, *Calamintha*, *Scutellaria*, *Nepeta*, *Polygonum*, *Rumex*, *Kochia*, *Atriplex*, *Arisaema*, *Scirpus*, *Carex*, etc.

Figure 49 shows the relation of the Pleistocene relicts of the Himalaya on the south Indian hills; as is evident, they are composed of the derivatives of both western and eastern floras. The figure shows further the major sources of the Himalayan floras. The Central Tien Shan, Pamirs, Western Tibet and the Northwest Himalaya constitute together the vast composite area of the Central Tien Shan High Ranges Province and the Pamir-Tibetan Pro-

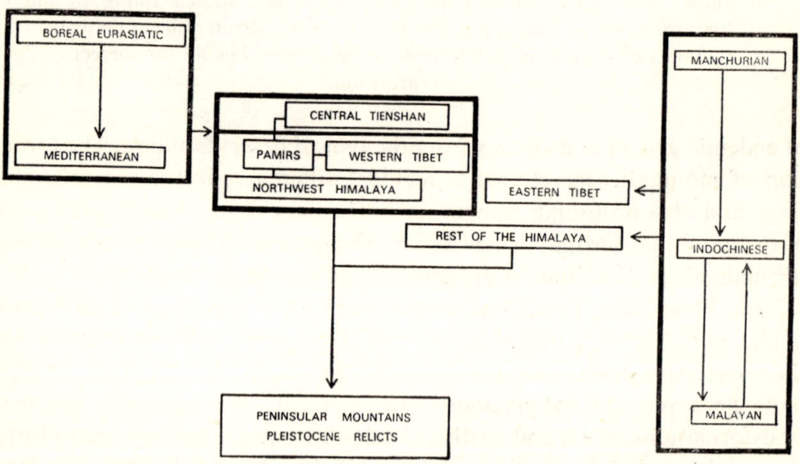

Fig. 49. Phytogeographic affinities and sources of floras. The Central Tien Shan, Pamirs, Western Tibet and the Northwest Himalaya constitute together the vast elevated areas of the Central Tien Shan Highland Province and the Pamiro-Tibetan High Ranges Province of Middle Asia (each shown separately in its own box). The major components of the high altitude flora of this complex region are derived from the Boreal Eurasiatic-Mediterranean complex (shown together as a separate box on the left). Eastern Tibet and the rest of the Himalaya, particularly the Eastern and Central Himalaya, are dominated by the components of the Manchurian-Indo-Chinese-Malayan floras (shown in one box on the right). The relation of Pleistocene relicts of the Himalayan flora on the high mountains of the south Indian Peninsular is also indicated. The arrows indicate the general direction of spread of floral components.

vince (shown in one major box) of Middle Asia. The major components of the flora of this region are derived from the Boreal Eurasiatic-Mediterranean complex (shown in a box of their own on the left). Eastern Tibet and the rest of the Himalaya, particularly the Eastern and Central Himalaya, are dominated by the components of the Manchurian-Indo-Chinese-Malayan floras (box on right). This figure is essentially a summary of the main conclusions on the affinities and sources of the high altitude flora of the Northwest Himalaya.

In the end it might be pointed out that the high altitude mountain vegetation of nearly the whole of Middle Asia, including naturally also the Northwest Himalaya as an integral part, is characterized by general xerophytization and ephemerization of the types, which had been formed previously in the region in the early phases of the uplift of the mountains. The nemoral mountain vegetation is generally reduced and the importance of the xerophile forms is increased considerably. As was shown to be the case with the high altitude insects of the region [75], the origin and development of the high altitude plant life of the Northwest Himalaya are integral phases of the Himalayan uplift. The high altitude vegetation seems to have become already well differentiated and nearly stable ecologically, when the insect life was still in the process of acquiring its high altitude characters. Recent and Pleistocene phases of the Himalayan uplift have contributed to the enrichment of the species of high altitude insects, but in the case of high altitude plants we observe a general change in the distributional pattern more than enrichment. Species of plants which had evolved at high altitudes of 3600–5000 m, having been subjected to xerophytization, are currently extending to higher elevations of aridity and colder conditions. The high altitude plants are therefore gradually ascending on the Himalaya at present.

7. SUMMARY

This chapter deals with a brief outline of the general floristic composition and geographical affinities of the plants of the high altitude ecosystem of the Northwest Himalaya. Tentative conclusions on the sources and evolution of the high altitude plants are also drawn from the available data.

The chapter begins with a brief review of the general geographical distribution of the typical high altitude genera and species occurring in the Northwest Himalaya. The distribution of genera, reviewed under each family, though well known in literature, provides the necessary basis for the elaboration of certain basic facts and ideas in a subsequent part of the chapter.

This review is followed by a brief analysis of four major floristic components of the high altitude ecosystem of the Northwest Himalaya viz., the boreo-alpine, including the circumpolar plants; the Euro-siberian or the Trans-palearctic plants; the southern European or the Mediterranean and the Siberian-Mongolian or the Manchurian plants.

The first three components are nearly equally strongly developed in the high altitude ecosystem, each constituting about 25% of the total. The Siberian-Mongolian component is somewhat less numerous. There is besides a very small and perhaps mostly insignificant percentage of other categories, including endemics, southern and eastern derivatives (Gondwana and Indo-Chinese-Malayan floras). The general pattern of altitudinal and horizontal distribution of the four major components is outlined. These components are compared with and identified against those generally recognized by earlier authors.

Particular attention is drawn to the prevailing misconceptions regarding the phytogeographical affinities of the Himalaya. It is pointed out that while the separation of the Western Himalayan Province from the Eastern Himalayan Province is basically sound, the boundaries between these two areas have been mostly confused by nearly all earlier authors. Their affinities have also been wholly misunderstood. The Western Himalayan Province of earlier authors is really an integral part *not* of India but of the Pamiro-Tibetan Province of Middle Asia; it is considered best as indeed only a district of that Province.

The high altitude ecosystem of the Northwest Himalaya presents a very pronounced affinity to the Alai-Pamirs and the Central Tien Shan. Between 85 and 90% of the genera and species of plants of the high altitude ecosystem of the Northwest Himalaya are common to the Alai-Pamirs. The ecological types, plant associations and formations met with at high altitudes in the Northwest Himalaya are also identical with those known from the Pamirs. About 75% of the genera and species from the high altitude ecosystem of the Northwest Himalaya are common to the Central Tien Shan. The high altitude plants, which are common to the Pamirs, Central Tien Shan and the Northwest Himalaya are listed. The Northwest Himalaya is described as biogeographically the southern fringe of Middle Asia rather than as part of India.

The phytogeographical divisions of Middle Asia are outlined. Ideas on the development of the flora of Middle Asia are developed. The sources of the flora, the uplift of the Himalayan System and its relation to the development of the arid steppes and the rôle of the Pleistocene glaciations are briefly mentioned. The high altitude flora of the Pamirs and the Northwest Himalaya is derived mainly from the Arcto-Tertiary forest floras of northern latitudes and from the ancient Mediterranean floras of savannahs from the south. The formation of the subarid-frigid steppes character is attributed to the Himalayan uplift.

Bibliography

1. Aitchison, J.E. 1888–1894. The Botany of the Afghan Delimitation Commission. *Trans. Linn. Soc. London*, (2) 3.
2. Anderson, T. 1863. On the flora of Bihar and the Mountain Parasnath, with a list of species collected by Messrs Hooker, Edgeworth, Thomson and Anderson. *J. Asiatic Soc. Bengal*, 32: 189–218.
3. Antevs, E. 1929. Maps of Pleistocene glaciations. *Bull. Soc. geol. America*, 40: 631–720.
4. Aymonin, G.G. & R.K. Gupta, 1955. Études sur les formations végétales et leur succession altitudinale dans las principeaux massifs du systéme alpine occidentale. Essai du comparaisson avec l'Himalaya. *Adansonia* (NS) 5: 49–94.
5. Baehni, C.C., E.B. Bonner & S. Vauthier, 1952. Plantes recoltées par le Dr. Wyss-Dunant au cours de l'expedition Suisse à l'Himalaya en 1949. *Candollea*, 13: 213–236.
6. Banerji, M.L. 1955. Botanical Explorations in East Nepal. *J. Bombay nat. Hist. Soc.*, 55: 243–268.
7. Banerji, M.L. 1963. Outline of Nepal phytogeography. *Vegetation*, 11 (5–6): 288–296, fig. 2.
8. Billings, W.D. 1975. Vegetation: Arctic and alpine Vegetation. Plant adaptations to cold summer climate. In: Arctic and Alpine Environments. Ed. J.D. Ives & R.G. Barry. London: Methuen & Co.
9. Bishop, B.C. *et al.*, 1966. Solar radiation measurements in the high Himalaya (Everest Region). *J. appl. Meteorol.*, 5: 94–104.
10. Blasco, F. 1971. Oreophytes of south India and Himalaya. *J. Indian bot. Soc.*, 50 (4).
11. Braun-Blanquet, J. 1923. Über die Genesis der Alpenflora. *Verh. naturf. Ges. Basle*, 25: 243–261.
12. Braun–Blanquet, J. & H. Jinny, 1926. Vegetations-entwicklung und Bodenbildung in der alpinen Stufe der Centralalpen. *Denkschr. schweiz. naturf. Ges.*, 43: 183–349.
13. Brockmann-Jerosch, H. 1919. Baumgrenze und Klimacharakter. *Beitr. geobot. Landschaft Zürich*, 6: 255.
14. Bukharin, P.D. 1961. Carotene levels of introduced plants grown at

various altitudes. *Dokl. Akad. Nauk SSSR* (Bot.) 136 (1–6): 28–30.

15. Burkill, I.H. 1910. Notes from a journey to Nepal. *Rec. bot. Surv. India*, 4: 59–140.
16. Burrard, S.G. & H.H. Hayden, 1907–1908. A Sketch of the Geography and Geology of the Himalaya Mountains and Tibet. Calcutta: 1–230.
17. Chandler, M.E.J. (Miss) 1933. The effects of the southern extension of glaciers and ice sheets on the Pre-Glacial vegetation. The nature of the flora as revealed by the plant remains associated with glacial and interglacial deposits. *Proc. R. Soc. London*, 118: 205–210.
18. Chatterjee, D. 1940. Studies on the endemic flora of India and Burma. *J.R. Asiatic Soc. Bengal*, (Sci.) 5: 1–69.
19. Christ, H. 1867. Über die Verbreitung der Pflanzen der alpinen Region der europäischen Alpenkette. *Denkschr. schw. naturf. Ges.*, 22.
20. Clarke, C.B. 1898. On the sub-subareas of British India, illustrated by detailed distribution of the Cyperaceae in that Empire. *J. Linn. Soc. London*, (Bot.) 34: 1–146.
21. Collett, H. 1921. Flora Simlensis. Calcutta.
22. de Terra. A. & T.T. Peterson, 1919. Studies on the Ice Age in India. Carnegie Inst. Washington DC 49: 354.
23. de Terra, H. 1936. Himalayan and Alpine orogenesis. *XVI Internat. geol. Congr.*, 2: 859–871.
24. Dey, A.C., M.R. Unniyal & V. Shankar, 1969. Flora of the Bhillaganga Valley of the erstwhile Tehri-Garhwal State. *J. Bombay nat. Soc.*, 65: 384–407.
25. Dobremez, J.F. & C. Jest, 1971. Cashiers Népalais Carte Écologiques du Népal. I. Région Annapurna-Dhaulagiri. CNRS Paris.
26. Dobremez, J.F. & C. Jest, 1972. Cashiers Népalais Carte Écologiques du Népal. II. Région Jiri-Thodung. CNRS Paris.
27. Dudgeon, W. & L.A. Kenover, 1925. The ecology of Tehri-Garhwal. A contribution to the ecology of the Western Himalaya. *J. Indian bot. Soc.*, 4: 233–284.
28. Duthie, J.R. 1893. Report on a botanical tour in Kashmir. *Rec. bot. Surv. India*, (1) 1: 1–18; continued in *ibid.* 3: 25–47 (1894).
29. Emberger, L. 1930. La végétation de la région Mediterranéene. Essai d'une classification des grouppements végétaux. *Rev. gen. bot.*, 42: 641–662, 705–721.
30. Engler, A. 1914. Über Herkunft, Alter und Verbreitung extremer xerothermer Pflanzen. *Sitzb. k.p. Akad. Wiss.*, 20.
31. Fedchenko, B.A. 1902. Ocherk rastitelnosti Pamirs, Shugnana i Alaiya. *Trudi S-Peterburg Obsch. Estetv.*, 33.
32. Fedchenko, B.A. 1912. Rastitelnost Alai i Pamira.
33. Fedchenko, B.A. 1927. Vertikalnoe rasprostranenie paporotnikov

Turkestankoi gornoi strani. *Izv. Glabnovo Bot. Sada*, 24.
34. Fedchenko, B.A. 1947. Botaniko-geographicheskoe raionirovanie Pamira, Tezisi i 45-lestiyu nauchnoi deytelnosti.
35. Fedchenko, O. 1903–1909. Flore du Pamir. *Acta horti Petropolitana*, 21 (1903) (Suppl.) 1, 24 (1904); 2, 24 (1905); 3, 28 (1907); 4, 28 (1909).
36. Flint, R.F. 1947. Glacial Geology and Pleistocene Epoch. New York, pp. 1–589.
37. Fyson, P.F. 1932. The Flora of the south Indian Hill Stations. Madras: Government Press. 1: 1–697; 2: 1–611 (plates).
38. Gamble, J.S. 1915–1935. Flora of the Presidency of Madras. London: Allard & Son Ltd. Parts 1–10, pp. i-lxiv, 1–2017 (Completed by C.E. C. Fischer).
39. Gams, H. 1923. Die Waldklimate der schweizeralpen, ihre Darstellung und ihre Geschichte. *Verh. naturf. Ges. Basel*, 35 (1): 262–276.
40. Gams, H. 1935. Pflanzenleben des Glockenergebietes, kurze Erläuterung der Vegetationskarte. *Z. deutsch-österr. Alpenver.*, 66: 157–176.
41. Gansser, A. 1967. Geology of the Himalaya. London: John Wiley Interscience, pp. 1–300.
42. Gaussen, H. 1933. Geographie des Plantes. Paris.
43. Goloskolov, V.P. 1949. Flora i rastitelnost visokogornikh poysov Zailiskovo Ala-Tau. Alma-Ata.
44. Golovkova, A.G. 1957. Visokogornaya rastitelnost Centralnovo Tian-Shanya. Sbornik Kirghizkovo Nauchno Issledovatelskovo Instituto Pedagogicki. Frunze.
45. Golovkova, A.G. 1959. Rastitelnost Centralnovo Tian-Shanya. Kirghiz-Gosudarst. Universitet Kafedra Bot. Frunze, pp. 1–456, fig. 128.
46. Golovkova, A.G. 1962. Geobotanicheskoe Raionirovanie Centralnovo Tian-Shanya. Kirghiz-Gosudarst. Universitet. Kafedra Bot. Frunze, pp. 1–139.
47. Good, R. 1964. The Geography of the Flowering Plants. New York. (3rd ed.).
48. Gupta, R.K. 1955. Botanical Exploration in the Bhillaganga Valley of the erstwhile Tehri-Garhwal State. *J. Bombay nat. Hist. Soc.*, 53: 581–594; 54: 878–886 (1957); 59: 486–512 (1962).
49. Gupta, R.K. 1962. Some observations on the plants of south India hill tops (Nilgiri and Palni Plateaus) and their distribution in the Himalaya. *J. Indian bot. Soc.*, 41 (1): 1–15.
50. Gupta, R.K. 1964. Bioclimatic types of the Western Himalaya and their analogous types towards the mountain chains of the Alps and the Pyrenees. *Indian For.*, 90 (8): 551–577.
51. Gupta, R.K. 1966. Bibliography of Himalayan ecology, forestry and phytosociology. *Excerp. Bot. Soc.*, (B) 7: 60–74.
52. Hara, H. 1966. The Flora of Eastern Himalaya. Tokyo.

53. Haviland, Maud D. 1926. Forest, steppe and tundra. Cambridge Univ. Press, pp. 1–466.

54. Hayden, H.H. 1907. Notes on certain glaciers in Northwest Kashmir. *Rec. geol. Surv. India*, 25: 127–137.

55. Heim, A. 1956. The geological structure of the Himalaya compared with the Alps. *Proc. nat. Inst. Sci. India*, 22 (A): 228–235.

56. Heim, A. & A. Gansser, 1939. Central Himalaya, geological observations of the Swiss Expedition 1936. *Denkschr. schweiz. naturf. Ges.*, 73 (1).

57. Hopkins, A.D. 1921. Bioclimatic zones of the Continents; with the proposed designations and classifications. *J. Washington Acad. Sci.*, 11: 227–229.

58. Hooker, J.D. *et al.* 1875–1897. Flora of British India. London: L. Breeve & Co. Ltd. Vol. 1: 1–740 (1875); 2: 1–792 (1879); 3: 1–712 (1882); 4: 1–780 (1885); 5: 1–910 (1885); 6: 1–792 (1894); 7: 1–842 (1897).

59. Hooker, J.D. 1891. Himalayan Journals. London.

60. Kanai, H. 1966. Phytogeography of Eastern Himalaya with special reference to the relationship between Himalaya and Japan. (*vide* Hara, H. 1966: 13–38).

60 a. Kashkarov, D.N., A.I. Zhukov & K.V. Stannyukovich, 1937. Kholodnaya Pustinya Centralnovo Tian-Shanya.

61. Kashyap, S.R. 1932. Vegetation of western Himalaya and western Tibet in relation to their climates. *J. Indian bot. Soc.*, 4: 327–334.

62. Kashyap, S.R. 1932. Some aspects of the alpine vegetation of the Himalaya and Tibet. *Proc. XIX Indian Sci. Congr. Bangalore*, 13–53.

63. Kingdon-Ward, F. 1960. Pilgrimage for Plants. London.

64. Kihara, H. 1955. Fauna and flora of Nepal Himalaya. Scientific results of the Japanese Himalayan Expedition to the Nepal Himalaya. *Kyoto Univ. Fauna & Flora Res. Soc. Kyoto*, 1–390.

65. Köppen, W. 1919. Baumgrenze und Lufttemperatur. *Petermans Mitt.*, 1919: 201–203.

66. Köppen, W. 1936. Das geographische System der Klimate. Handb. Klimatologie 1 (C): 1–44. Berlin: Gebrüder Bornträger.

67. Korovin, E.P. 1934. Vaprosam palaeo-ecologicheskikh smien v Srednii Asii. Vaprosi Ecologii i biocaenologii.

68. Korovin, E.P. 1934. Rastitelnost Srednii Asii i Yuzhnovo Kazakhstana. Tashkent. Vol. 1.

69. Korovin, E.P. 1962. Rastitelnost Srednii Asii i Yuzhnovo Kazakhstana. Tashkent Acad. Nauk UZSSR. Vol. 2.

70. Krishnan M.S. 1973. Geology. In: Mani, Ecology and Biogeography in India. The Hague: Dr. W. Junk, Monographiae Biologicae 23: 60–97.

71. Landon, P. 1928. Flora of Nepal. London: Constable & Co. Vols. 2.
72. Larsen, J.A. 1975. Timberline: Ecology of the northern continental forest border. In: Arctic and Alpine Environments. Ed. J.D. Ives & R.G. Barry. London: Methuen & Co.
73. Löchert, H. 1969. Pflanzen an Grenzstandorten. Stuttgart.
74. Lundegårdh, R. 1930. Klima und Boden in ihrer Wirkung auf das Pflanzenleben. Jena (2nd ed.) pp. 1–480.
75. Mani, M.S. 1962. Introduction to High Altitude Entomology. London: Methuen & Co. pp. 305.
76. Mani, M.S. 1968. Ecology and Biogeography of High Altitude Insects. The Hague: Dr. W. Junk BV Publishers, pp. 1–524.
77. Mani, M.S. 1973. Fundamentals of High Altitude Biology. New Delhi: Oxford & IBH. pp. 1–196.
78. Mani, M.S. 1974. Ecology and Biogeography in India. The Hague: Dr. W. Junk BV Publishers. Monographiae Biologicae 23: 1–727.
79. Marquand, C.V.B. 1929. The botanical collections made by Capt. Kingdon-Ward in the Eastern Himalaya and Tibet in 1924–25. *J. Linn. Soc. London*, 48: 149–229.
80. Meher-Homji, V.M. 1972. Himalayan Plants on south Indian Hills: Role of Pleistocene glaciation versus long distance dispersal. *Sci. & Culture*, 38: 8–12.
81. Meher-Homji, V.M. 1973. A phytoclimatic approach to the problem of Mediterraneity in the Indo-Pakistan Sub-Continent. *Feddes Rep.*, 83 (9–10): 757–788.
82. Meher-Homji, V.M. 1974. Disjunct distribution in plant kingdom. *Sci. & Culture*, 40: 217–227.
83. Meher-Homji, V.M. & K.C. Misra. Phytogeography of the Indian Sub-Continent. In: Progress of Plant Ecology in India. Ed. R. Misra *et al.* Vol. 1, pp. 9–88.
84. Merriam, C. 1898. Life zones and crop zones in the United States. *US Depart. Agric. Biol. Surv. Bull.*, 10.
85. Meusel, H. 1971. Mediterranean elements in the flora and vegetation of the West Himalayas. Plant Life of the South-West Asia. *Bot. Soc. Edinburgh*, 53–72.
86. Meusel, H. & R. Schubert, 1971. Beiträge zur Pflanzengeographie des West Himalayas. 1. Teil: Die Arealtypen. *Flora*, 160: 137–194. Parts II and III, pp. 373–432 and 573–606 respectively.
87. Nair, N.C. 1964. On a botanical tour to Lahaul and Spiti (Punjab Himalaya). *Bull. bot. Surv. India*, 6: 219–235.
88. Nakao, S. 1964. Living Himalayan Flowers. Tokyo.
89. Numata, M. 1966. Vegetation and Conservation in Eastern Nepal. *J. College Arts Sci. Chibo Univ.*, (Nat. Sci.) 4 (4): 1–18.
90. Numata, M. 1967. Notes on botanical trip in Eastern Nepal. I.

J. College Arts Sci. Chibo Univ., (Nat. Sci.) 5 (1): 57–74; II. *ibid.*, 5 (2): 243–258 (1968).

91. Paulsen, O. 1920. The studies in the vegetation of Pamir. Copenhagen.
92. Popov, M.B. 1926. Ecologicheskie typi pustyn Srednei Asii. *Izv. Glav. Bot. Sada*, 25.
93. Popov, M.B. 1927. Osnovnye cherty razvitiya flory Srednei Azii. Byull. SAGU.
94. Popova, I.A. 1960. A pigmentakh listev pamiriskikh rastitelnosti. *Bot. Zhur.*, 43 (11): 1550–1561.
95. Raikova, I.A. 1945. Opit botanikogeographicheskovo obzora Pamira. Tezici Dokla. Nauchnoi Cessi 25 CAGU. Tashkent.
96. Raikova, I.A. 1945. K istorii formirovaniya rastitelnovo pokorova Pamira v svyzi formirovaniem obshego relefa i relefa dolinnikh Mestoobitanii. Trudi CAGU (NS) 1. Trudi Central. Asiat. Gosud. Universitet.
97. Rao, T.A. 1959. A botanical tour to Milam Glacier. *Bull. bot. Surv. India*, 1: 97–120.
98. Rao, T.A. 1960. A botanical tour to Pindari Glacier and Kumaon Hill Stations. *Bull. bot. Surv. India*, 6: 47–57.
99. Rau, M.A. 1960. On a collection of plants from Lahaul. *Rec. bot. Surv. India*, 2: 45–56.
100. Rau, M.A. 1963. Illustrations of West Himalayan Flowering Plants. Calcutta: Botanical Survey of India Special Pub.
101. Rau, M.A. 1973. Flora and vegetation of the Himalaya. In: Mani, Ecology and Biogeography in India. The Hague: Dr. W. Junk BV Publishers, Monographiae Biologicae 23: 247–280.
102. Royle, J.E. 1839–1840. Illustrations of the botany and other branches of natural history of the Himalayan Mountains and of the flora of Cashmere. London.
103. Schröter, C. 1908. Die Pflanzenleben der Alpen. Zürich: A. Rausten.
104. Schweinfurth, U. 1957. Die horizontale und vertikale Verbreitung der Vegetation in Himalaya. *Bonn. geogr. Abh.*, i-xii, 1–373, maps 2.
105. Shaw, C.H. 1909. Vegetation and altitude. *Plant World*, 12 (3): 63-64; *Amer. Nat.*, 43: 420–421 (1909).
106. Shaw, C.H. 1909. Causes of timberline on mountains. *Plant World*, 12 (8): 169–181.
107. Smith, W.W. 1913. The alpine and subalpine vegetation of southeast Sikkim. *Rec. bot. Surv. India*, 4: 323–431.
108. Smythe, F.S. 1938. The Valley of Flowers. London.
109. Stannyukovitch, K.V. 1948. Botanikogeographicheskoe raionirovanie vostachnovo Pamira. Soobshena Tadj. FA Nauk SSSR, 5.
110. Swan, L.W. 1961. The ecology of the High Himalaya. *Sci. Amer.*, 205 (4): 68–75.

111. Swan, L.W. 1963. Ecology of the heights. *Natural History*, 23–29.
112. Swan, L.W. 1967. Alpine and Aeolian regions of the world. In: Wright, H.E. Jr. & W.H. Osborn, Arctic and Alpine Environments. Indiana Univ. Press, 3: 29–54.
113. Takhtajan, A. 1969. Flowering Plants: Origin and Dispersal. Edinburgh: Oliver & Boyd.
114. Tansley, A.G. 1935. The use and abuse of vegetational concepts and terms. *Ecology*, 16: 284–307.
115. Tilman, H.W. 1952. Nepal Himalaya. Cambridge Univ. Press, pp. x-271, pls. 61.
116. Troll, C. 1959. Die tropische Gebirge, ihre dreidimensionale klimatische und pflanzengeographische Zonierung. *Bonn. geogr. Abh.*, 25.
117. Troll, C. 1967. Die klimatische and vegetations-geographische Gliederung des Himalaya-Systems. Ergebn. Forsch.-Unternehmen Nepal Himalaya, 1 (5): 353–388.
118. van Steenis, C.G.G.J. 1934–1936. The origin of the Malaysian mountain flora. *Bull. Jardin Bot. Buitenzorg*, (3) 13: 135–262, 287–417; 14: 56–72; 1962. The mountain flora of the Malaysian Tropics. *Endeavour*, 173–189.
119. Wadia, D.N. 1931. The syntaxis of the Northwest Himalaya, its rocks, tectonics and orogeny. *Quart. J. geol. min. Soc. India*, 65: 189–220.
120. Walker, H. & E.H. Pascoe, 1907. Notes on certain glaciers in Lahaul. *Rec. geol. Surv. India*, 25: 139–157, pl. 40–46, 60, 62.
121. Wardle, P. 1975. Alpine timberline. In: Arctic and Alpine Environment. Ed. J.D. Ives & R.G. Barry. London: Methuen & Co.
122. Webster, G.L. 1961. The altitudinal limits of vascular plants. *Ecology*, 42: 587–590.
123. Zimmermann, A. 1953. The highest plants in the world. Mountain World, New York: Harper, pp. 130–136.
124. Zimmermann, A. 1956. Resultates des expéditions scientifique genevoises au Népal en 1952 et 1954. Partia Botanique. *Candollea*, 15: 127–147.

Index